MW00817547

# ENRICHING ANIMAL LIVES

**Hal Markowitz**

**Illustrations by**

**Jenny Markowitz**

Mauka Press

Pacifica, California

Maukapress.com

ISBN 978-0-9833579-1-9

*For Krista, Jenny, Wendy, Tanner & Tim*

# Table of Contents

# 1

# Introduction

This is largely a recipe book. Good fortune has led me to many parts of the world, and colleagues have been exceptionally kind in asking advice about how to mount effective programs in enrichment. In these discussions, a pervasive comment by my hosts has been a wish that my earlier publications had included more extensive detail about "how to do it" for animals in their care. Consequently the majority of the material in this new book focuses on a wide variety of animals and some suggestions for how environments might be enriched.

I met my dear friend and colleague Terry Maple when he was a graduate student at the University of California, Davis, and I visited to consult at their primate center. In the decades that have passed since then, Terry has had considerable influence on the zoo world. I was thinking about writing a more general book about my philosophy of behavioral enrichment, but during one of a number of trips to Zoo Atlanta where Terry was Director I mentioned this to him. Those of you who know Terry will confirm that he is a large man both in size and influence on the world of zoo biology. His *instruction* to me was that, "You *must* write a cookbook to tell us how to make equipment that will increase naturalistic behavioral opportunities for zoo animals." So, if this book is too slanted toward being instructional and disproportionately full of recipes for potentially increasing opportunities for zoo animals......blame Terry!

I have further blame to share for my inadequacies. Some of you will find my attempts at humor in this book to be insipid, rude, or not to your taste. Please blame more than a half century of students and other audiences who had the bad judgment to stay awake more regularly when I inflicted my sense of humor on them as part of my lectures.

Finally, I am too cowardly to accept total blame that this book contains a lot of segments that are probably excessively autobiographical, dwelling on my experiential reasons for suggestions about the construction and use of enrichment equipment. Sprinkled throughout the text are thoughts about my own philosophy of science and how to enrich your own life as well as lives of others. Please blame all those class evaluations in which I was told (anonymously, and thus I trust honestly) that students felt their learning experience was enhanced by having a teacher incorporate personal experiences and honest personal feelings about the subject material.

Looking at my earlier text *Behavioral Enrichment in the Zoo* while making plans for this one, I marveled at how far we had come conceptually, but how little we have *progressed* in terms of financial support for environmental enrichment. While understanding the need to provide more adequate stimulation for captive animals has become widespread, and there are international meetings dedicated to the topic, enrichment remains something of a nicety rather than a fundamental need in the eyes of many.

Perhaps most critical in this area is the fact that many administrators of institutions that house animals continue to adopt the most economical means to satisfy legal requirements for enrichment rather than seeking the most effective way to enrich animal lives. On the positive side, I have found an increasing number of folks in more local control of purse strings who work hard to divert as much money as they can to improve opportunities for captive animals. But it continues to disturb me that adequate budgets for significant empowerment of captive animals are not routinely provided. Greg Timmel and I have discussed some possible reasons for this in a recent book chapter (Markowitz & Timmel 2005).

I have intentionally avoided specifying exact components and schematics in most cases for three reasons. First, it is my firm belief that bright individuals undertaking efforts based on these ideas will undoubtedly come up with a myriad of improvements on the general concepts provided here. Secondly, it is a waste of time identifying suppliers for specific components since by the time readers choose to undertake any of the proposed projects many of the best component sources will have changed. Thirdly, it has been my experience that when local technicians are involved in the final design of equipment they will be better able to keep it serviceable. However, one must be extremely careful in selection of equipment designers and construction personnel (see Chapter 20).

The limitations of budgets and how that affects apparatus choice will be addressed in Chapter 20. It brings me real pleasure when I see clever colleagues find inexpensive ways to accomplish measurably useful enrichment equipment for animals in their care. Sometimes, if one is skillful and willing to spend hours beyond the call of duty, it is possible to make amazingly complex and efficient equipment for less than ten percent of what it would cost to buy prefabricated components.

***Sorry, sorry, sorry***

I apologize in advance for the fact that your favorite group of animals may not be the topic of attention anywhere within this book. It would obviously be impossible to include all types of animals (apologies to you parasitologists looking for enrichment for your critters!). My primary emphasis is on those megavertabrate groups where I have some experience and I hope that my suggestions will be stimulating for you. Readers will undoubtedly find many glaring omissions. Much more could be written about laboratory animals. Whole large groups of animals commonly used in research and occasionally as pets, such as lagomorphs, have been largely ignored here but not in our other published works (e.g., Markowitz & Timmel 2005; Sorrells et al 2003).

Along with the recipes I have interspersed a number of personal accounts of experiences working with various species. I hope that readers will not take offense if I sometimes seem excessive in emphasizing the need for caution in enrichment efforts on behalf of exotic animals. During my years of work in this area I have seen friends lose lives or become seriously and permanently disabled when they were working with potentially dangerous animals. My only motive lies in the hope that my caveats may help reduce the frequency of such tragedies.

My final advance apology has to do with the fact that I recognize I have not been consistent in the style in which I have written chapters in the sense that most cookbooks are. In many cases it will be apparent to the reader, without my continual cross-referencing of chapters, that many of the recipes may easily be modified to work well for species beyond those used as examples. And in some cases I frankly just had more I wanted to share about some species with which I had worked.

***Where to begin enrichment efforts: By knowing the animals whose lives you would improve***

My first recommendation to students and other colleagues who ask my advice about how to design enrichment equipment for some exotic species is always the same. First, learn as much as you can about this species both in the wild and in captivity. Where budgets allow, there is nothing more invigorating than visiting them in their natural habitats and observing their behavior as unobtrusively as possible. When this is impossible, the modern world of internet resources makes a wealth of information available to anyone who truly wants to study almost any species. While field work and extensive literature searches require considerable effort, the rewards of this labor far outweigh the necessary work and occasional hardship.

The two happiest parts of my current research life are those spent in the field studying animals in their own habitats and those spent watching animals empowered by opportunities provided them in captivity. In preparing for our first major work in behavioral enrichment in the zoo, I had a great time studying everything that I could about gibbons and watching as much as I could find available in the way of motion pictures showing their behavior in nature. Even though I knew our budget would not allow anything approaching a natural habitat for these wonderful apes, I was inspired to think that there must be some affordable way to allow them to use their natural talent to gather their own food in captivity. I designed, constructed and installed equipment that allowed some gibbons in the Portland Zoo to learn to feed themselves when they wished to do so.

The first time I saw Harvey, a white handed gibbon feeding himself by brachiating between response areas to activate food delivery (Markowitz 1972) the effect on me was multiplicatively more than I expected. It made me feel guilty about the vast number of animals that we humans keep powerless in captive circumstances. The gibbons in this cage no longer had to wait for keepers to drop food on the floor and sit awkwardly to procure their rations. They had incentive to exercise and use their natural talents to obtain a healthy variety of food chunks high up in their environment. At the same time, it gave me hope that we might, in a small way, make the lives of captive animals more stimulating and naturalistic. Twenty five years later when I watched Sabrina, a beautiful old black leopard (*Panthera pardus*), cavorting like a youngster chasing an acoustic "bird" up and down a tree segment in her cage I still found it thrilling (Markowitz et al 1995).

For readers whose immediate response is that you would rather see exhibited animals engaged in fully natural behavior in truly natural environments, this is a topic that will be discussed extensively in the second chapter. Captive facilities cannot generally provide anything approaching natural environments for wild animals. Dangers to animals, including humans, obviously preclude the struggles for survival in the wild such as predation on other species and lethal battles with conspecifics. Regulatory inspectors, visitors and supporters of captive animal facilities would act to close places that exposed animals to the risks and death rates evident in nature.

**Evaluating the utility of enrichment efforts**

In the next chapter, some of the critical questions to be asked when considering new enrichment ventures are addressed. No matter what your well-defined goal may be, it is important that you provide evidence of how the work that you have accomplished has enriched the lives of the animals in your care. This involves the very difficult task of conveying to those who review your work what exactly your definition of enrichment involves, and then providing clear and extensive evidence showing the extent to which your enrichment aim was reached. Unfortunately, far too many of us frequently fail in this important area. It is not sufficient to simply show that an animal engages her/himself with something that you have proclaimed as enriching. After all, we animals all frequently engage ourselves in things that are more enriching for others than they are for us. It is possible for us to quit our jobs if this becomes intolerable and we can afford it. Animals in captivity do not have that luxury. Consequently, we need to be vigilant in efforts to see that they can *choose* to use enrichment opportunities that we provide.

For example, if we define enrichment as "the opportunity to engage in species typical behavior," we need to provide evidence that the animal uses new "opportunities" in a willing and long-term manner. If our definition is that "intra-species aggression due to captivity will be reduced," this must carefully be quantified both before and after enrichment efforts are made. Of course, if all that you mean by "environmental enrichment" is that you have improved the appearance of the environment, then you might as well evaluate the response of the visitors to the environment rather than pretending without evidence that the captive animals are better off.

When enrichment efforts take place in institutions devoted to exhibiting animals for public viewing, it is always a good idea to measure visitor response to your efforts. Where we have done this in the past, we have invariably found that visitors spend much more time at exhibits where they can observe animals which are active because they have been empowered to control some aspect of their environment than at similar exhibits lacking these qualities. Visitors' spoken and written responses also show much more respect for animals when they observe their abilities and cleverness in making the most of these empowerments.

**About "Enrichment"**

Sometimes what you name a book can greatly influence the popularity of a work. At least that is what the publishers of my first single author book told me many years ago. I had a working title of "Behavioral Engineering in the Zoo." Some critics of our work, who had never gotten past the term "behavioral engineering" in the titles of earlier articles and presentations, suggested that we were making robots of animals by engineering their behavior. Those who actually read or listened beyond the titles knew that we referred to engineering *environments* that would provide greater opportunities for resident animals.

I had a very bright editor from Van Nostrand Reinhold who was familiar with this controversy and was consequently delighted that I had changed the title to *Behavioral Enrichment in the Zoo.* He predicted that I would be happy with the changes brought by this renaming. To my surprise, when the book appeared I received many letters from folks congratulating me for the "new work" that we were doing and telling me how much better they liked this than our old attempts to "control animals." Of course the book simply summarized the "old work."

The shift to "enrichment" as the focal term has proven to be of value in a critically important way. It allows readers to better understand the intent of efforts in which we try to provide environments that give more power to others. Providing lives that are richer for captive animals may not be easy, but it may ultimately be the most important thing we can do in any of the capacities where we are responsible for the care of others.

You will find that I often use "we" and "our" throughout this book in referring to previous work. This is not to aggrandize myself, but rather to connote that I could not have accomplished this work without the help of wonderful colleagues, too numerous to cite in every case. I also recognize that grammarians and many colleagues will object to my frequent use of personal pronouns when referring to animals, but this is not an unintentional mistake in usage. It is because I think of animals as individuals and am more comfortable talking about them that way.

# 2

## Philosophical Issues

**Who?** **Why?** **Where?** **What?** **How?** **$$$$$?** **When?** **Evaluation?**

### Who?

Choice of species whose captive lives you wish to improve probably "came naturally" for you. But, on the chance that you have not thought about the animals most in need of enrichment, let's take some time to explore some issues.

Watching zoo visitors can be an educational experience, both interesting and perplexing. Listening to their complaints about the plight of animals in captivity is an exceptionally worthwhile exercise in refining our own thoughts about many of the issues to be covered in this chapter. Many people who visit zoos have not had the luxury of visiting any of these exotic animals in their native habitats. Only a small percentage of zoo visitors spend much time watching nature shows that are *authentic* or visiting the library or internet to study carefully about nature. It is not surprising that many of their evaluations miss the mark. Folks who have seen lions (*Panthera leo*) in Africa without invading their space with a noisy motorized vehicle know that they should expect to see lions asleep during much of the day. But visitors to zoos typically stand in front of lion exhibits trying to entice these cats to be active or complaining that "all they do is sleep."

Zoos get constant complaints about wolves (*Canis lupis*) that "are pacing unnaturally because they are bored." These complaints are not forthcoming from those who have carefully studied wolves' behavior in the field. Redundant movement back and forth across limited routes is such a common part of the repertoire of wolves in the wild that hunters stalking them with cameras or guns use well-worn tracks to find them. In selecting a focal species for our efforts in enrichment it is critical that we think beyond how attractive they are to visitors. We need also to evaluate what might be truly measurably "enriching" for a species and whether there is a practical means to work towards the innovations that would produce a better life for these animals.

Such evaluations will lead one to the often discussed parallel question about whether "enrichment" should be a synonym for "more like nature" (Coe 1989; Markowitz, 1982; Markowitz & Eckert 2005). What leads to a better life for captive animals may not be the same as what spices the lives of the same species in the wild. For example, providing captive predators an opportunity to hunt for other megavertebrates may make the lives of their caregivers less safe and may lead to less opportunity for loving treatment by their keepers. Indeed, it can be argued that those characteristics that make a predator most suitable for reintroduction to the wild may in many cases be the obverse of those that make it most suitable for captive display and care (Hediger 1955). It behooves us to carefully evaluate what kinds of goals we have for the animals on which our enrichment efforts will focus. If captivity is temporary and we hope to return these animals or their offspring to nature, our goals may be quite different than if we plan for them to make permanent residence in the zoo.

The Association of Zoos and Aquariums (AZA) has adopted as one of its goals the use of zoos as "arks" in returning endangered species to the wild. Consequently some readers may wish to focus their efforts on ways to enrich the captive environment that may help to prepare their species for potential life in natural habitats. However, I offer as a caveat that in cases where there have been many generations of breeding and socialization exclusively in captivity, we should not expect that the genes of these animals will make them most suited for survival in nature. The conservation of behavior requires the continued preservation of some of those environmental qualities that provide the selective pressures that have resulted in the specialized nature of successful animals in the wild (Markowitz, 1998).

**Why?**

At first it may seem obvious that we wish to enrich the lives of animals in captivity because we love them and believe that we can make their lives happier and more comfortable. But there are a number of more subtle questions about our motives that deserve attention. Let's explore some of those considerations.

Why are these animals in captivity and what are the long term intentions for them? The first possibility is that they are in captivity because they are your pets and they give you comfort. In this case, your goals will be far removed, and in many senses simpler, than if the animals are temporarily captive and intended for reintroduction to native habitats. By and large, you will pretty much decide what to do depending on what appears to you empathetically to give them most joy. Most folks feel that they are so bonded to their companion animals that it is clear to them when these animals are happy or sad. It would be foolish for me to try and argue with this, and probably lead many readers to close and condemn this book! But, I would suggest that studying as much as you can about the species and observing conspecifics may provide recipes for enrichment beyond those engendered by empathic ability. Further, it may be helpful to consider carefully the extent to which you wish to empower your companion and the extent to which you desire to maintain control over this animal. As discussed below, these two motives are often at odds with each other.

Perhaps you have decided to improve the environmental stimulation for a species on display in a zoo or other recreational/educational facility. Then, your first and most over-riding questions concern why you wish to enrich the environment. Is it in order to make these captive representatives of their species more compelling arguments for the conservation of these animals in nature? Or is your focus on making life more stimulating and fulfilling for your focal animals? Sometimes it is possible to address both of these goals at once, but at other times there is clear conflict.

If, for example, you believe that the primary mission is to enhance the role of these "emissaries" for their species by attracting attention to their behavioral beauty, you may focus on getting as many visitors as possible to see a predator successfully capture prey. This will almost certainly distort visitors' impressions of nature, since predators in nature are notoriously unsuccessful in the majority of their efforts (Cheney, 1978). However, watching a leopard pounce upon a free flying bird and bring it down is a much more memorable experience for most casual observers than watching it unsuccessfully stalking some potential meal. In these days of time-lapse TV presentations and overzealous "experts" who present their "daring" stories, most people have developed quite faulty impressions about animal life in the wild. It is certainly arguable that, in order to compete with these media for attention to important conservation causes, it is defensible to do whatever most successfully attracts people to your educational mission.

My own decision, and that of many colleagues who have worked closely with me, has been to try and make life more stimulating and species-appropriate for captive animals, and to secondarily focus on combining this with public education about nature wherever possible. It has been our good fortune to find that when captive animals are provided more power to voluntarily engage in naturalistic behaviors, they typically choose to do so and are much more attractive to visitors (Markowitz 1982; Markowitz et al 1995; Markowitz & Eckert, 2005).

There is also the possibility that you want to improve the lot of animals maintained in some research facility. In light of increasing federal requirements for attending to the "psychological well-being" of research animals, there has been particular attention given to increasing enrichment efforts. Unfortunately, there has seldom been equal attention to providing the budgets necessary to do a truly good job in these efforts. We have proposed that making environments richer in terms of species-typical opportunities may make these captive animals more adequate "models" for the evaluation of experimental manipulations (Markowitz & Timmel 2005). One change that is sometimes feasible is housing non-human primates that are naturally social in compatible groups rather than individually.

This may enhance our abilities to predict whether the effects of drugs on these primates are likely to be similar to the effects on social humans in clinical trials.

This is obviously a lot to consider. I will not complicate matters with the dozens of other "why" questions that come to mind. Instead, it is left for you to decide why you wish to do this work, and your careful consideration will logically affect your answers to the questions that comprise the rest of this chapter.

**Where?**

Having decided on a species that you wish to work with, you have undoubtedly been thinking about where you would like to study them, and *where you can afford to study them.* (A typical response: Africa, *the library and the zoo.*) Regardless of what your budget may dictate, it would be silly *not* to spend a good deal of the time reading the work of those who have extensively studied your species of choice. One obvious reason is the fact that you may spend many hours in the field for just a few minutes of observation of some of the megavertebrates. For example, in seven years of research on black howler monkeys (*Alouatta pigra*) in Belize, my students and I identified jaguar (*Panthera onca)* tracks a few times, but only once did we observe a jaguar directly in the field.

One great advantage of learning about the behavior of animals in their natural habitats, whether you are able to do that first hand or are limited to studying the work of others, is that it may allow you to identify some of the important natural contingencies that help to shape their behavior. In this manner, it may be possible to identify potential ways to provide them some power in their captive circumstance. For example, you may find that an important part of their lives involves stalking some particular prey. Perhaps you can devise a way to allow them the joy of stalking and capturing prey in captivity. Of course, you will have many potential problems using live prey should you choose to do your work in public areas of a zoo. One of the reasons that zoos cannot be truly natural is because many of the visitors who provide the revenue that keeps zoos afloat will not tolerate seeing any living creatures preyed upon by other animals. In some senses this is understandable because most exhibits cannot allow prey species to escape as they would most often do in the wild. While I can sympathize with that view, there are other views that are more difficult to entertain, such as the proposal that even carnivores in captivity should be taught to be vegetarians. (Honest, I did not make that up....I have received such requests!)

Whatever your feelings on controversial matters such as these, your careful observation and resource study will undoubtedly lead you to the conclusion that it would certainly be nice to have the captive environment encompass more of nature than it typically does. Perhaps you may even conclude that if we could just bring a little part of Africa to the zoo so the lion could be seen in its own ecology, all would be well. This brings us to a great dilemma facing zoos. Since the vast majority of captive animal facilities are constantly struggling for the money necessary to continue to attract paying guests and donors, it is critical that they make their institutions attractive to visitors. Many architectural firms that focus on zoos have whole teams of architects and media specialists who are prepared to develop glossy plans to attract monies for zoos to build new exhibits. It is much easier, for example, for a zoo to get a bond issue passed if they can have TV spots that show pictures of a wonderful plan to bring part of Africa to the zoo. The dilemma lies with the word "part" in the previous sentence, because the part that is most frequently missing involves the natural contingencies which give vitality to the focal species in the wild. For example, predators without prey, and vice versa, do not make for an active exhibit or for honest opportunity to view nature.

Sometimes zoos have spent fortunes building habitats to look natural only to find visitors complaining because they "can't see the animals." Then the zoos are forced to make renovations such as removing foliage that provides privacy for the animals, or revamping the exhibit to provide feeding areas where provisioned animals can be viewed. The contrast between this and nature is clearly apparent.

Some people have historically tried to contrast the behavioral enrichment approach with a more naturalistic one in which one simply brings real nature to the captive animal and assume that the captive animal's nature will be released. This is, of course, a false dichotomy because none of those who have worked in our group have ever objected to making exhibits appear more like nature. Certainly, it is more rewarding to see an animal on a natural substrate and with real trees and bushes than it is to see them in the concrete jungles that have served as zoos historically. Given infinite budgets it would be wonderful if all captive creatures could have such surroundings. But with limited budgets, public unwillingness to see some parts of nature in captivity, and legal and ethical mandates against exposing animals to the dangers that typify natural environments, there are tough decisions to be made. Many times decisions come down to whether to spend the available dollars primarily on making the exhibit more responsive and entertaining for the residents, or making the physical environment look more attractive to paying customers.

Since our emphasis is consistently on improving the *measurable* well-being of animals, if a choice needs to be made we would prefer environments that promote the animals' well-being over those that appear attractive or natural to humans. However, this choice is only made necessary by budgetary constraints and/or bad facility design (Forthman-Quick 1984; Foster-Turley & Markowitz 1982; Markowitz & Gavazzi 1995).

No matter what your thoughts on these philosophical matters and what your decision about where to study your focal species, it is critical that you make the most informed choices possible before you commit your limited time and resources to any enrichment plan. Study, study, study, wherever you can. When you have devised a plan and implemented it, evaluate the extent to which you have succeeded in providing the animal(s) a means to respond to some naturalistic contingencies, and to what extent this opportunity is attractive to the animals that you wish to benefit.

**What?**

Now you are confronted with the difficult task of deciding things like: how complex should your enrichment project be? For some animals such as hoof stock the project may be very simple and inexpensive, such as seeing that food is routinely distributed throughout the environment in a way that reduces aggression and allows browsing without confrontation. On the other end of the spectrum, apes and dolphins may require a great deal of complexity if the enrichment is to provide them continuing stimulation. Even with animals capable of significantly complex responses, the nature of your enhancement to their environment will depend in part on what is already there. If an arboreal primate such as a gibbon has no place to exercise above ground, a first step will almost certainly be to provide some opportunity for brachiating.

Assuming that you have an environment that provides the minimum appropriate equipment for your animal of choice, and assuming that you wish to give this creature more control of some aspect of its environment, it is now time to use the expertise that your study and observation have provided. Can you design something that will really be attractive to this animal and be used as long as it remains available to them?

Perhaps I can be of help by providing a couple of starting points based upon many years of mulling over these questions. If we are interested in improving the well-being of an animal, it is important to know that there have been a number of studies illustrating that responsive environments may help animals to combat negative behavioral and physiological symptoms of stress. For example, in some of our own studies we have seen that primates given control to deliver food and music to themselves whenever they wished bounced back much more quickly from some of the stresses of captivity (such as being restrained for physical exams) than did matched counterparts with no such opportunity (Line et al 1989; Markowitz & Line 1989). Carefully quantified results like these, and my good fortune for decades in being able to witness the differences in the behavioral vigor of animals whose behavior brings reinforcing results as compared with those that are limited to toying with inanimate objects (such as their own feces) have shaped my conclusion: You can have the greatest positive effect by providing animals a means to control some aspect(s) of their environment.

It has long been established that many animals prefer to work to control delivery of their own food, even when that food is available in plain sight and can be obtained without effort. Allen Neuringer first made this phenomenon popularly known in quantitative work by publishing his well controlled laboratory studies in operant conditioning (Neuringer 1969). I know that this is more than some kind of conditioning anomaly because I have observed exactly the same phenomenon in a wide variety of megavertebrates where we provided some parallel opportunities (Markowitz 1982; Markowitz & Eckert 2005; Markowitz & Timmel 2005). I insert this information here to reassure you that it will not be necessary to "deprive" animals of food or water in allowing them to exercise control in this fashion. However, many captive animals including some of us humans might benefit from less ad lib feeding.

Some readers will be disturbed by the use of conditioning terminology in parts of this book, just as some colleagues have misunderstood some of our early work and thought that it was some kind of "Pavlovian" plot. This may, then, be an important place for us to consider the difference between: 1) analyzing what sorts of contingencies occur in nature for a particular species, in order to construct analogous contingencies for captive members of the species; and 2) forcing animals to perform in the manner that we wish. Envisioning classical or Pavlovian conditioning, most folks immediately think about something they have read, e.g., about restrained animals who hear some signal and then have meat powder placed on their tongue to make them salivate, eventually causing them to salivate to the previously neutral signal. Operant (or instrumental) conditioning is better characterized by a cartoon which I like to share with classes when I am teaching about the distinction. In this cartoon two rats are in a cage and one says to the other "Boy, do I have this experimenter conditioned. Every time I press the lever he gives me food!"

While I have introduced this distinction in what some readers may see as a trivial manner, I believe that there are some important reasons for those interested in behavioral enrichment to examine some of the philosophical considerations about these matters in some detail. I have repeatedly suggested that one should examine natural contingencies in depth before deciding on a means of behavioral enrichment. This is because the behavioral repertoire of animals is a joint function of heritable natural proclivities and the life history of an individual. The behavioral communalities shared by particular members of a species result from many similarities in their contemporary environments *and* similar genetic makeup of these individuals.

If in the course of their evolutionary history there has been selection resulting in a species that feeds high in trees and tends to limit itself to fruits and leaves, this is most probably because ancestral relatives of the individuals that exhibited such responses were more likely to survive. For example, animals that continued to dwell on the ground or seek food at ground level might be more likely to be consumed by predators. We often find that arboreal animals are very selective in the kinds of fruits and leaves that they consume, and this may be a consequence of contemporary bad experience with other foods that prove poisonous, or it may be an inherited tendency based on which ancestors have survived.

The extent to which behavior is heritable and the extent to which it is based on the life history of individuals is a topic of much conjecture that is beyond the scope of this book. But, it is essential for one's understanding of much that will follow that contingent control of behavior is not something unique to the laboratory. What we ourselves continue to do is largely dependent on what happened the last time that we did it. The same is true for other animals. Finding what there is to respond to in the environment and seeing what happens when one does is a significant part of how all animals, including humans, function.

**How?**

The first and easiest path that you may choose in your enrichment efforts might be for you to buy a pre-manufactured enrichment device from one of the many suppliers of such equipment. If you choose this alternative it is important to avoid purchasing equipment before making thorough evaluation, especially if you intend to purchase more than one such item. Thorough evaluation means several weeks of observing the utility of the device in producing measurable results for your animals. Ensuring that animals do not cease use of the device once it is no longer novel is especially important in the case of devices that are essentially toys and have no rewarding characteristics contingent on the animals' behavior. If the device is complex and constructed of multiple parts, it is important to ensure that animals cannot easily disassemble the equipment and consume parts or use them as weapons.

My research group at the California Regional Primate Research Center at Davis conducted one of the few published carefully quantified studies of simple toy use by captive primates. During routine observation we had noted that many of the toys that were placed in the cages of older monkeys got little attention after the first day or two. Our study confirmed that passive enrichment toys did not alter the behavior of aged rhesus monkeys beyond a very short period (Line, Morgan & Markowitz 1991).

After reading about some of the past work which has used complex electronics and/or computer controlled devices for enrichment you may decide that you wish to design an environment with multiple responsive opportunities for your animal. In chapter 20 you will find many cautions concerning undertaking this unless you have ample time and considerable familiarity with such equipment. I encourage you to read both some of the anecdotal examples and the accounts of false starts in chapter 20, which concentrates on apparatus.

In any case, it is essential that you get to know those who take regular care of the animals that you choose to work with, and to ensure that they are really devoted to the continued maintenance and use of the equipment. Many times zoo, aquarium, or wild life park personnel who are very excited by the attention that comes with innovation are less excited about the labor required to keep these important opportunities for animals in regular use. "Novelty effects" is terminology often used to refer to the fact that animals are frequently more responsive to what is new and unique. This applies as well to humans responsible for attention to the chores of maintaining enrichment protocol.

**$$$$$**

Institutional budgets for enrichment are not typically very large. Consequently, unless you are independently wealthy you may have to take on projects that are not too ambitious or seek outside funding. Whether you seek funding from the place where you accomplish the enrichment or from donors or grants, your work will have greater chance for support if you spend sufficient time in the planning stages and produce a preliminary document that illustrates your knowledge of the species. Your detailed plan and request for funding should show that you have good knowledge about the focal species in estimating the probability that the scheme will be attractive to the animals. Finally, there should be a detailed budget that does not underestimate the real cost of producing, installing, and maintaining the equipment involved.

A description of the systematic methods of observation that will be used to evaluate changes in the animal's behavioral and/or physiological well-being may also enhance your chances of obtaining financial help for your work. In consulting previous articles on enrichment, you may be disappointed to discover, as I have been, that the majority of these published works do not report much in the way of systematic evaluation. For those readers unfamiliar with methodology for systematic observation there are many resources that are available such as Phil Lehner's *Handbook of Ethological Methods* (Lehner, 1979, 1996).

**When?**

Opportunities for immediate funding or encouragement of your plans by animal keepers or other friends may lead you to wish to "strike while the iron is hot." But it is critically important not to begin your effort without reasonable study, thought and planning of the sort that has been emphasized in this chapter. It is more important to develop a reputation for producing enrichment schemes that work and keeping promises with respect to when they can be adequately produced than it is to get a foot in the door.

**Evaluation?**

The need for careful systematic evaluation of the measurable benefits for animals has already been stressed. However, there is another critical aspect of evaluation that will determine your future potential as an enrichment worker. That is how others evaluate your productivity and the usefulness of what you have accomplished for their animals and their institution. Sad to say, you will find that this is a tricky business because once you depart from the place in which you have accomplished your work, you will seldom have a say in its use and maintenance.

To show how difficult a problem this may become, I will close with a personal account of one of our own great disappointments in the use and consequently the long-term evaluation of some of the best enrichment equipment with which my research crew has ever been involved. By staying up late at night and devoting considerable parts of our own homes as well as our lab space to producing naturalistic equipment for a tropical rainforest zoo, my coworkers and I were able to make equipment for a number of animals that made us proud. It was sophisticated, safe, good-looking, durable, and made to withstand the humid environment in which it was installed.

In one case, computerized equipment solicited visitors to join in by choosing the order in which hunting activities would become available for tigers (*Panthera tigris*). Great care and effort had gone into production and we spent a summer installing the equipment and ensuring the animals' familiarity with its use. But we were certain that it would not be routinely used after we left unless there was someone on staff whose duty it was to work with this equipment and who had been trained to do any necessary repairs or maintenance. Consequently, before the final contract was signed, I also required written assurance that one of the people that we had trained would be added to the zoo staff. The highly competent person we chose was experienced in building and maintaining equipment and in working with animals. He moved in with my family and me for many months so that he could participate in the construction phases and learn exactly how to care for the equipment. Despite the legal contract, shortly after we departed the county that funded the zoo declared that they were bankrupt and the written promises were not honored. Years later when I revisited this zoo a number of times I found the equipment in surprisingly good shape in spite of the fact that it was not in use.

I was further saddened by hearing from colleagues who had read about the work and traveled distances to visit the rainforest zoo and were told that nothing was working because we were not there to make it work. Whatever you can do to help ensure that your enrichment efforts will not go to waste or be improperly evaluated will be time well spent.

Of course, given the opportunity to do this all over again, I would doubtless accept the opportunity. Seeing the enjoyment that the animals had, if all too briefly, was wonderful and the learning experience for all of us was terrific. Students who worked with us on the project still tell me how much fun it was and what a sense of accomplishment they had from participation.

# 3

# FELINES

Beginning here, many chapters will include considerable design details. I recognize that the more technical suggestions and specifications may be of little interest to readers who want only general ideas and rationale for various kinds of enrichment. Since this is intended in large part to be a recipe book, like any good cook you will want to skim or ignore those parts of the cookbook in which you have no interest. And of course cooks always love to spice recipes to their own taste.

Cats are wonderful animals with which to work for any number of reasons including their independence, cleverness in solving problems, and individuality. In a number of felid species we have found that adult females and the young are more likely to engage in active pursuit of new enrichment opportunities. Males are inclined to engage in those familiar behaviors of sleeping, showing off to conspecifics, and trying to share their sperm with as many females as possible. In our enrichment efforts for most feline species we have found the females more responsive and willing to work. Perhaps this is because we have never offered sexual opportunities as incentives!

## Passive Enrichment

Anyone who has had a cat for a pet can suggest dozens of things that they will play with when these items are first introduced or when a human is there to share the activity. As with all of the passive reinforcement we have tried, habituation is a factor here and frequently exchanging balls, things to chew on, or other passive enrichment equipment is highly advisable.

Where it is feasible, encouraging social interaction among cats may be enriching. Visual, olfactory, and acoustic stimulation have also been used in attempts to enrich feline environments. We have found that these sensory stimuli produce much more consistent and enduring increases in healthy activity when they are part of enrichment programs providing cats a chance to pursue simulated prey. This is especially true if the cats' pursuit behavior sometimes leads to a chance to eat as it might with real prey in nature (Markowitz & LaForse 1987).

Schuett and Frase, in an article titled, "Making scents: Using the olfactory senses for lion enrichment" (2001), share one of the few quantitative reports of the behavioral outcomes of presenting various scented substances to felines, although a number of zoos I have visited used scents extensively to provide stimulation for big cats.

McPhee has conducted an empirical study of the effects of whole carcass feeding as enrichment for large felids, both on and off exhibit (McPhee 2002). While her study does illustrate limited efficacy for this approach, McPhee acknowledges that with a larger sample the statistical results might illustrate even more promise for this technique. I have seen whole carcass feeding generate extended bouts of species-typical behavior in European zoos such as the one in Copenhagen where it is routinely employed. Infrequent use of this productive technique in U.S. zoos is largely a function of concern about public reaction to witnessing that the foods consumed were not "made at the supermarket" but are recognizable parts of once living creatures.

## Simple Active Enrichment Devices

Saskia and Schmid (2002) examined "The effect of feeding boxes on the behavior of stereotyping Amur tigers" in the Zurich Zoo. The method employed feeding boxes with electromagnetic locking mechanisms distributed throughout the exhibit. These boxes were filled with meat early each morning and strong electromagnets were turned on closing the access doors to the boxes. Periodically throughout the day semi-randomly selected boxes had the magnet turned off, thus allowing the tigers to gain access by opening the sliding doors. The article includes a nice diagram of the apparatus designed for this study. It would be only slightly more complex to design a simple computer program or dedicated hardware to automate the whole enrichment procedure. That would relieve staff of the periodic requirement to return throughout the day once the feeding boxes were loaded. The results reported by Saskia and Schmid were complex and showed more positive effects for the female tiger involved. Their general conclusion was that this was an effective method in reducing some stereotypic behavior that occurred in these tigers as a function of captivity. They also indicated that an important addition might be the use of an acoustic signaling device to let the tigers know when these opportunities were available, e.g., as suggested in Markowitz and LaForse (1987).

In the pre-testing stages before seeking approval and funding for an automated responsive enrichment system for servals (*Leptailurus serval*) Shirley LaForse and I designed a labor intensive means to assess the interest of servals in chasing artificial prey. While readily giving approval for our preliminary work because it provided interesting activities for some docents and other volunteers, the zoo director said to me: "You know, of course, that cats quickly habituate to things like this and will only show interest for a few days." Months later the female serval continued to show very active interest in the opportunity for this very simple artificial prey "capture." Shirley and I smiled silently when the director brought along a visitor and told them how "we" had all anticipated that this active opportunity would be of continuing interest to these cats. If you have a limited budget but lots of person power this is a system that was easy to build and of demonstrable value in enrichment:

### Recipe 1. Simple artificial prey chase

1. Measure the longest straight open pathway on the cage exhibit substrate and obtain transparent 4" diameter cast acrylic tubing with .125-inch walls sufficient to span most of this dimension.

2. Conceal the ends of the tubing in hollowed out tree stumps at the cage perimeter.

3. Purchase a durable stuffed toy rodent and attach each end securely to lengths of the strongest line that you can find. Place the rodent and line inside tubing.

4. Attach a concealed clothesline retractor to one end of the line so that the rodent will automatically move back to that end of the cage when tension on the line is released.

5. Arrange a safe means for food such as meat balls to be dropped into the cage near one of the tree stumps. With most exhibits this can be accomplished by simply passing a piece of sturdy tubing through the perimeter fence and angling it downwards so that food will easily drop through.

6. In each tree stump install a simple water resistant speaker and run speaker cords (outside the cats' enclosure) to a place near the string pulling station. Route the speaker cords through toggle switches so that they can be connected manually to a sound source such as a voice synthesizer or digital playback device.

Trial and error testing is necessary to see what works best in terms of scheduling opportunities for a hunt. The simplest general procedure is as follows. At semi-random times turn on the sound of a squeaking rodent at the tree stump where the artificial prey is currently hidden. If the cat forages at that stump within your pre-selected maximum duration for the prey sound, have the human operating the system move the rodent swiftly across the cage. Pouncing on the tubing directly over the rodent that is "scurrying" through it represents capture, and a food treat should then be delivered. If the cat fails to forage at the appropriate stump during the time that the rodent squeak is coming from it, or forages but then fails to pounce on the prey, begin the semi-random time interval before the next hunting opportunity will occur.

Figure 3-1. Serval About to Pounce on Artificial Prey Emerging from Tree Stump

For the female serval, this simple manually operated hunting opportunity was so rewarding that she virtually never missed an opportunity to respond (figure 3-1). Each afternoon large meaty bones were presented in the back room of this barren exhibit so that the animals could enter an area where they would not be bothered by the public during their feeding activity. If the female heard the squeaking sound of her "prey" she would bring the bone outside and alertly listen and watch for clues that she might have a chance to forage, pounce on the rodent, and gain some food treat. Anecdotally, I noticed some occasions when she did not even immediately consume the food treats, but just anxiously awaited a chance to "attack" the only moving prey that she could expect to regularly encounter in this captive environment.

## Recipes for Automated Responsive Enrichment

### Recipe 1. "Souping up" the feeding boxes described by Saskia and Schmid

The best and ultimately cheapest alternative for this recipe is to install an appropriate input/output (I/O) interface board in a computer. These boards are relatively inexpensive and any competent computer technician can help you select the correct kind of board depending on your I/O requirements. For this recipe you will want a board with output components capable of driving solid state relays that can handle the power required to reliably operate the electromagnetic locks. There must also be some outputs with sufficient sensitivity to handle audio level signals with minimum degradation. The inputs may be simple switch closures or significant voltage swings depending on your choice of peripheral equipment. Most I/O boards are equipped to handle these kinds of input and output requirements very reliably and to protect the computer and other components by use of onboard opto-isolators. However, if your computer I/O connection does not provide complete isolation for remote components that produce high voltage electrical noise on activation or deactivation (such as the door-locking mechanisms employed in this recipe) you can ensure complete electrical separation of the computer from high power currents by using external solid-state electronic relays with complete opto-isolation.

The control and data recording functions are so simple for this enrichment paradigm that very little of a computers' memory and only a small part of the I/O capabilities of your interface will be used. Consequently it is very practical to use a single computer to control the workings of enrichment such as this in a number of exhibits. There are readily available software packages and compact discs (CDs) that will allow you to choose a variety of prey sounds and to control the duration for which they are heard through the speaker that has been selected for a particular foraging opportunity. One can also change the sound for each occurrence thus better simulating changes that would occur in nature. This should prove more stimulating for both the cats and viewers than redundantly presenting the same prey sound.

1. Using materials for construction strong enough to withstand the felines' strength and dexterity, make a series of boxes or other containers for the food. Locking devices for the food doors can be manufactured by using readily available solenoids and the working parts of a large brass or stainless exterior door lock. A quicker and cheaper alternative for cats with less massive strength than lions is to purchase a mechanism that allows remote control of exterior doors. There are some surprisingly rugged and reliable electromechanical devices that are used for controlling steel outside entry doors in neighborhoods that are prone to having home robberies. If you have big bucks or lots of talented staff to help in construction of exhibit components, you can make these "boxes" into any desirable form depending on the environment that you wish to design for the species. For example, they could have the appearance of a small mound in the terrain or a tree where the animal forages for food. The only important precautions are that whatever food you use cannot be easily broken or granulated so that it

may potentially impede the sliding door, and that the sliding door must move in tracks that will not catch the claws of the cats.

2. Install a magnetic switch to detect the fact that the animal has opened the door. These switches can be purchased for less than ten dollars because they are commonly used in security systems. They consist of two separate components: a permanent magnet to be attached to the door itself, and a magnetically operable reed switch encased in a protective coating. The builder will need to ensure that these components are protected from destruction by the animals. This should not be too difficult since the permanent magnet requires no electrical or other connections, and the switch will be located on the non-moving door housing where it can be enclosed in a small box with a conduit connector so that the leads can be routed to the control mechanism.

3. The door opening detector provides the opportunity for many desirable enhancements in the enrichment paradigm. These include easy automatic monitoring of the animals entry so that this event and its time can be recorded, and the possibility of automatically closing the doors and reloading the boxes with meat to provide multiple feeding opportunities at each station. If you choose this more complex alternative, then you could avoid the use of magnetic locking devices and instead use a motor to control opening and firmly closing each compartment. While initially more expensive, motor controlled doors would be more reliable and have longer service lives.

4. Just inside the top of each box, build a baffle that protects it from the cats, and install a speaker in this protected area. Run the speaker wire for this box so that it is protected from destruction by the residents. The wires coming from the door locking mechanism and the door opening detector also require protection from damage by animals. Depending on local electrical code requirements, it may be convenient to run all the wires by the same safe route. The general recipe for routing in the next paragraph may be easily modified for your specific needs. In some designs, where it was totally impractical to safely route wires over long distances or irregular terrain, we have used infrared or radio controlled relays at remote locations rather than directly running wires to the control apparatus.

5. Install very strong and well anchored electrical conduit between the boxes and one piece of conduit to the control equipment from the nearest feeding box. Conceal the conduit in some fashion by arranging plants around it or by some other means depending on the substrate. If possible, place the conduit beneath the surface of the substrate. For example, by digging a deep narrow trench in which to bury the conduit and then refilling the trench, you can make a more attractive appearance. But remember that some of these species have great digging abilities and a penchant to use them. Run all of the speaker leads and electrical cords for the door mechanisms through the conduit to a common place where they can be routed to the enrichment controller.

If this is to be a permanent rather than a temporary experimental change in the environment, there will be special legal requirements for the kind of wire and conduit that you employ. For temporary situations we have found it convenient to simply use inexpensive zip-cord of the proper gauge required for safe transmission based on the power level and distances to be traversed. Two or three conductor cord is available in any hardware or electrical store and costs only pennies a foot when purchased in quantity. Having separate sets of zip-cord for each device *and remembering to carefully label them* makes it easy to connect them to the appropriate parts of the control and recording mechanisms. A color-coded multiconductor cable may also be a good choice, but will take longer to install and may not be legal if the electrical code requires separate bundles for high and low voltage wires. Time-consuming labor will result from the need to remove the cable sheath at several points in order to connect to the components housed in various locations.

6. There are two logical choices for storing attractive sounds for playback. The first is to simply store sounds on the hard drive of your computer. Alternatively, you can record prey sounds on a separate digital chip for each box so that distinctive prey species can be acoustically represented in each location.

7. Connect the wires coming from the magnetic detector on each door to your control circuitry in a manner that triggers the program to turn off the sound and prevent future choice of this hunting location until it is restocked with food and the circuitry is re-enabled. Each designer will have their own favorite way to accomplish this. The easiest method may be to simply use the magnetic switch in series with the circuitry that produces sound for this particular location and enables it to be chosen as a hunting site. Then when the permanent magnet is moved away from the switch because the door is open, this part of the circuit will be non-functional. If this method is employed it will be necessary to ensure that the cat cannot re-close the door once it has been opened.

8. Connect the wires coming from each of the lock mechanisms to the I/O board. A simple computer program can disable the appropriate lock when it is selected as the active hunting site. If the cat does not open the box during the hunting time period that you have programmed, activate the lock preventing access to the box.

In summary of the procedure (figure 3-2): a cycle of hunting begins when a sound is produced at one of the box locations and the lock for this box is released; if the cat opens the box from which the prey sound is coming, access to the food is immediate, sound terminates, and a non-hunting period begins; if no hunting occurs during the hunting period that you have programmed, the sound ends, the door is relocked, and a non-hunting period begins; when your pre-selected quiet time ends, the procedure begins again at a randomly selected hunting site.

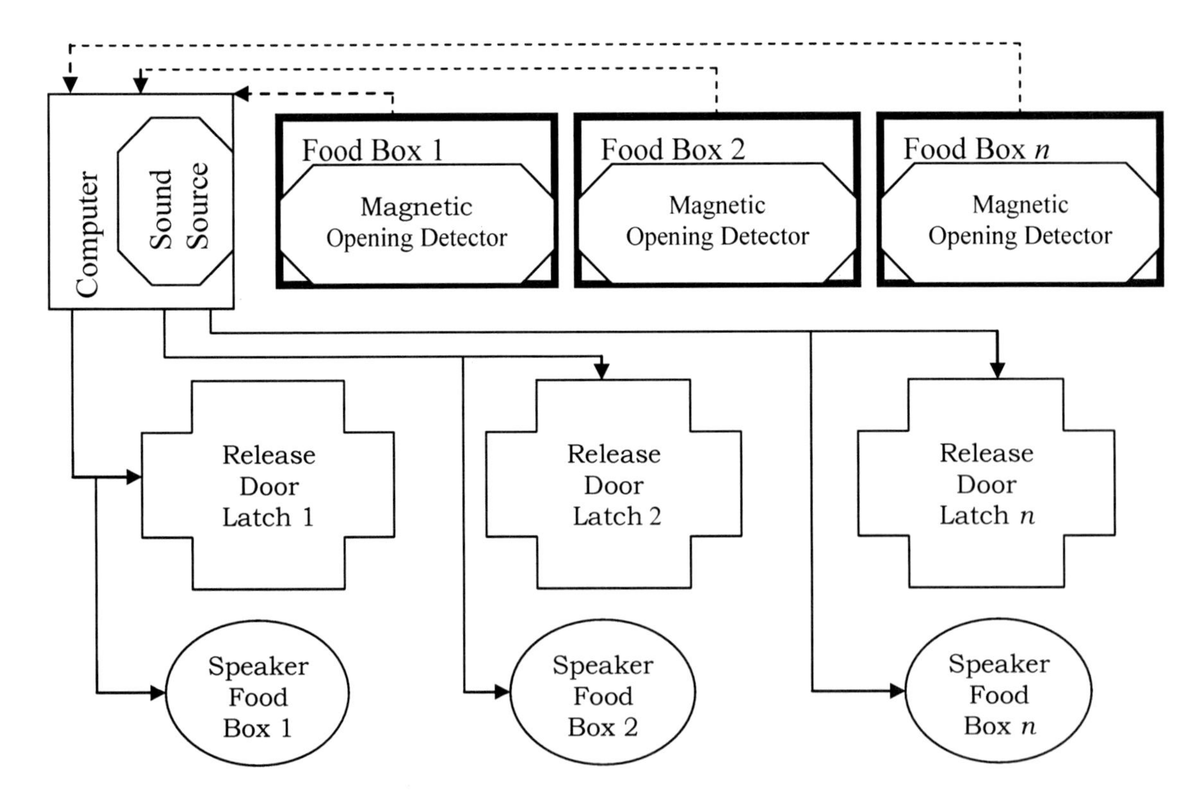


Figure 3-2. Component Arrangement for Automated Food Lockers

Since you are inventive and know the individual animals whose lives you hope to enrich, you will doubtless wish to add nuances of your own to this simple paradigm. For example, you might wish to elaborate the program to include a longer rest period when food has been obtained then when successful foraging has not occurred. You might decide to place all of the hunting stations at areas adjacent to the perimeter, thereby allowing you to use a food conveyer to automatically reload boxes that have been emptied, thus allowing more extensive hunting periods. If your purpose is to call the attention of visitors to the felines and stimulate healthful exercise and more frequent opportunities to hunt, you might use small portions of food for each successful hunt and run the program more frequently throughout the day. Since you can easily produce pseudo-random intervals with your computer, you might choose to have hunting opportunities occur at unpredictable times as well as random locations. I leave the fun of making these choices to you rather than further complicating this recipe.

**Recipes 2 & 3. Active Artificial Prey**

There are myriad ways to produce mobile artificial prey, but many of them are fraught with problems for application in zoo environments (See Chapter 20). Here are two simple recipes that may produce relatively safe and reliable moving artificial prey for felines if carefully followed. In both cases a first step is to select a captive habitat that has sturdy tree limbs that the cat can safely climb and leap onto, and that can easily be ascended without danger. Ideally these will be living trees. For more barren environments you may have to drag or crane in large dead tree segments as we have sometimes done in enrichment efforts.

Unless you have a capable electronics designer and technician, both of whom are willing to work without wages, you will find it cheaper and more versatile to use an inexpensive computer and I/O board for these recipes. These will be specified as controllers in the recipes for purposes of brevity.

**Recipe 2. Prey Moved Through Randomly Placed Transparent Tubes**

1. Arrange lengths of sturdy transparent tubing joined with elbows as needed to follow the path that you wish the prey to run. Make the lowest points in your path for the prey at each end of the tubing. Use the widest angle elbows possible in routing your path for ascending, descending, or turning.

2. Select a lightweight realistic-looking stuffed prey animal that fills most of the diameter of the tube but will not bind in any elbows in the path. It is important that you select an item with ambiguous front and back ends so that it will not look ridiculous when animated in a reverse direction. Test to make certain that the prey will not jam by passing it several times throughout the tubing by means of strings temporarily attached to both ends. If there is any resistance discard and replace or fix the artificial prey so that it moves freely and with only a very light pull on the strings. Once you have ensured that the prey moves freely in the tube, remove the strings.

3. Seal or cap the ends of the tubes and make threaded holes in each end so that high pressure air hoses can be coupled to them.

4. Bring the other end of each high pressure hose to the output side of a separate solenoid valve that is made to control air flow. Connect the input sides of the solenoid valves to a compressed air source.

5. Install a detector for movement near each end of each tube. Inexpensive options are to use a photo cell system or to modify a cheap driveway motion detector as described in chapter 20. This motion detector will be used to register the presence of either the artificial prey or the predator.

6. Install a mechanism appropriate to deliver food to the cat for successful hunts. The most reliable and versatile feeder that we have found can be constructed by obtaining a sufficient length of strong 35mm Mylar film leader and arranging it in an endless loop which is driven by a toothed stainless steel drive sprocket mounted on a motor shaft. Motors that are used in commercial food dispensing machines have proven extremely reliable and long lasting for this purpose. A general description of construction of this type device is provided below.

Prepare a length of straight hardwood with a smooth dadoed slot very slightly more than 35mm wide running its full length. The length of the hardwood track should be based on the size and number of food items you desire to be available for automatic delivery before reloading the belt. Many factors will have to be weighed in this decision, such as available space and whether the food is perishable. Carefully connect a motor bracket so that the motor driven sprocket will run just beyond the dadoed piece of hardwood in a position allowing the endless loop of film to run smoothly in this slot. At the far end beyond the hardwood film carrier, install a spring loaded device with a smooth 35mm small reel so that appropriate tension will be present to prevent the belt from coming off the drive sprocket at the other end.

It is very important that the motor drive, tension belt and dadoed track all be attached from one side in mounting them in an enclosure, and that the other side have sufficient space so that the belt itself can easily be removed by depressing the tension mechanism and slipping it off. This is necessary for regularly sanitizing the belt and replacing belts when they break (figure 3-3).

Finally, place a 3 or 4 inch diameter piece of tubing just beyond the end of the belt driven by the sprocket. When the feeder is advanced the selected distance for food dispensing, food will be scraped off the belt by the edge of the tubing and fall through the tube. For reliable dispensing, the tubing should be of rigid plastic and exactly vertical in installation. If it is necessary to route the food through a side opening in the enclosure rather than a top one, run the tubing vertically for most of its distance so that the deposited food will have sufficient velocity to carry it through a 45 degree elbow installed on the bottom of the tube and protruding into the enclosure. Length of belt movement desired for each food delivery can be easily adjusted by altering a timer setting or computer-generated time duration.

Figure 3-3. Feeder Belt Apparatus with Endless Loop of 35mm Mylar Film

7. Place a small computer with an appropriate I/O board in a protected place, and run connections to each of the solenoid valves, to the motion detectors for the animals, and to the feeder mechanism (figure 3-4).

8. Engage a competent programmer to arrange the sequence of events described in the next paragraph, and program automatic data recording. Useful and easily recorded data include time and location of animal responses, the time that prey were running, and the number of times that food was successfully earned as a function of "capture."

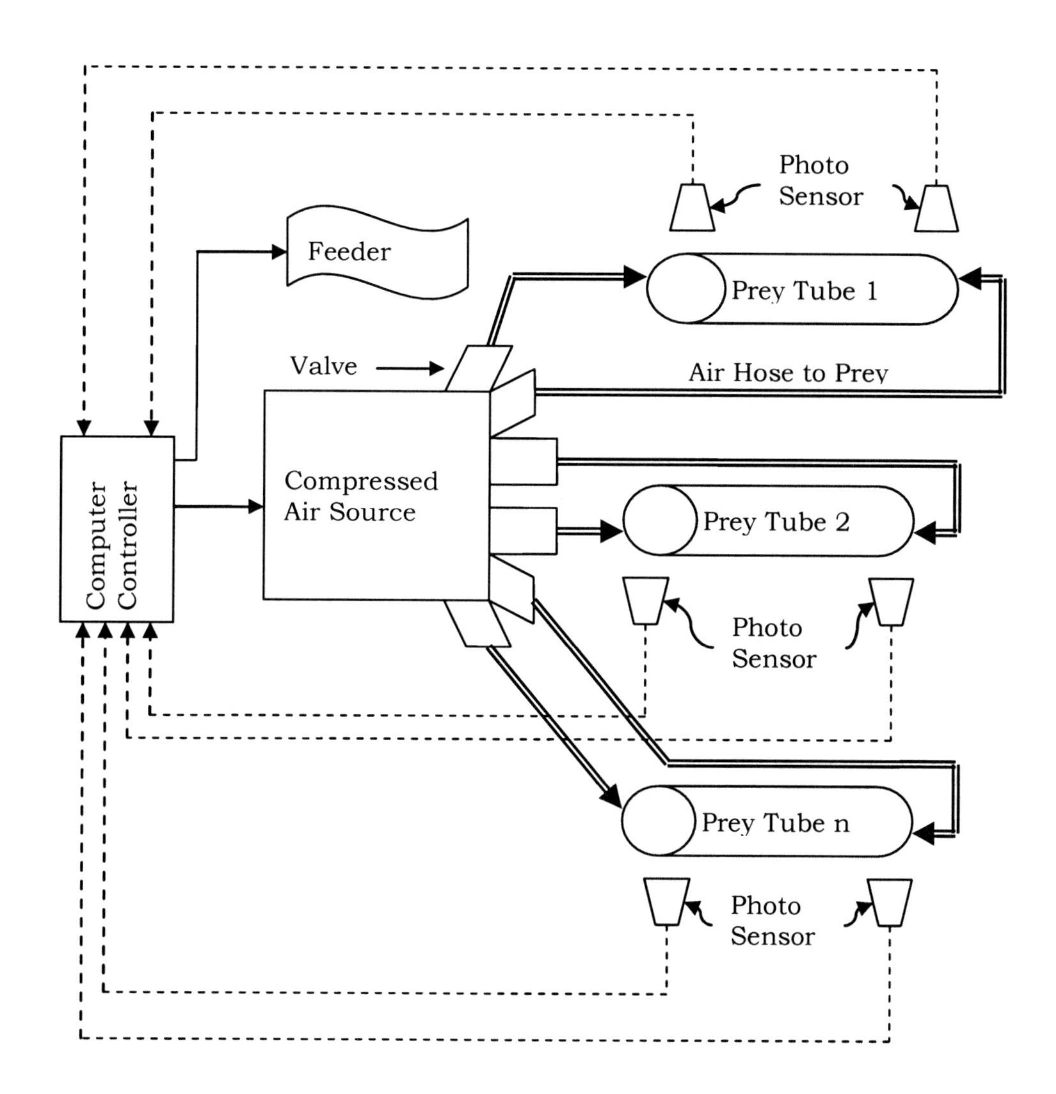


Figure 3-4. Block Diagram of Apparatus for Automated Tube Prey

An easily accomplished and useful addition to this recipe is the incorporation of sounds at the places from which prey will imminently leave. This can be done by installing waterproof speakers and running protected wires from these speakers to the I/O board to deliver computer generated sounds to the selected output.

Sequence of events: At random times within a range selected by the enrichment specialist (that's you!), randomly select one of the locations in which the prey rests at the end of the tube. Shoot air through the tube by opening the appropriate solenoid valve, thus forcing the prey to move through the tube until it reaches the opposite end. If you have added sound capability, turn on prey sounds at the starting end briefly before the prey begins its travel. If the predator reaches the terminal point at the same time as the prey, consider this a "capture" and deliver food. Whether or not a capture occurs, once the prey has run all the way through the tube generate another random time interval before the next potential hunt begins.

Do not forget that cats are very good at finding ways to outsmart enrichment workers! Be certain that your program includes ability to detect position of the cat when each chase begins. Otherwise the cat may learn to simply rest at one end of the prey's run until the prey arrives at that point where simultaneous presence of predator and prey results in treat delivery.

Of course, as with all my recipes, you may desire to modify and improve on this one. For example, you may wish to install more detectors for movement of predator and prey and allow "capture" to occur anywhere along the path of travel if predator and prey arrive there simultaneously. This would be highly desirable if you have extremely long runs of tubing for the prey.

**Recipe 3. Acoustic Prey**

In our more recent work in complex enrichment for felines we have found solely acoustic, apparently moving prey to be of remarkable utility (figure 3-5). The cats have shown continuing interest in pursuing the moving sounds a considerable portion of the times that they occurred.

Figure 3-5. Black Leopard Pursues Acoustic Prey by Climbing Tree with Concealed Speakers

This kind of system is much less difficult to construct and to control than one with moving component parts. It is also quite inexpensive by comparison. The simplicity of using acoustic prey is illustrated by the brevity of this recipe.

1. Program prey sounds onto solid state chips or produce them from the control computer as discussed in Chapter 20.

2. Place a sequential series of speakers at frequent intervals along tree limbs in a manner that protects them from destruction by animals or weather. Connect these speakers through cables in hidden conduit to the computer which is in a remote safe location.

3. Select an appropriate method for detecting movements of the predator based on the particular details of your apparatus location. Inexpensive commonly available gyro-type series of detectors can be arranged so that movement of specific tree limbs by the predator will jiggle the branch sufficiently to produce an output, indicating movement. The wiring for these detectors may be passed through the same conduit with speaker wires to the computer.

4. Select an appropriate kind of feeder and location for the feeding apparatus. Connect the feeder to an output on the I/O board

5. Have a competent programmer provide the following sequence and arrange for your required data to be automatically stored (figure 3-6).

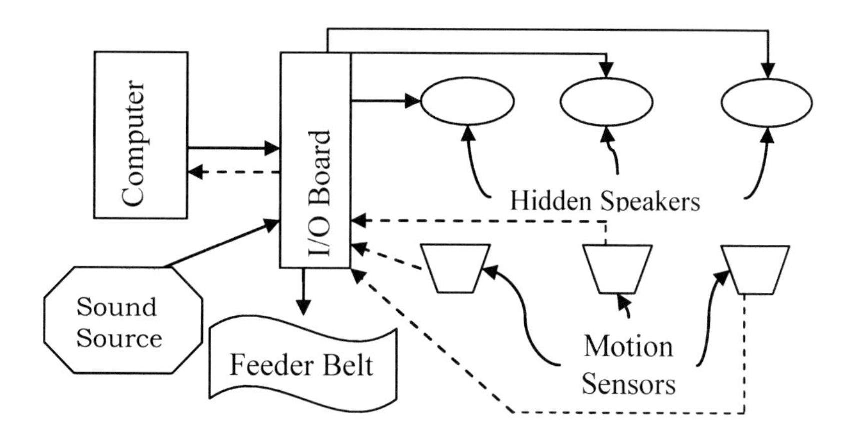


Figure 3-6. Block Diagram of Apparatus for Audio Chase

At random times within a range selected by the enrichment specialist, produce sequential movement of sounds through a randomly selected path of speakers. Ensure that this is a path along which prey could feasibly move. If the input signals from the motion detectors indicate that the predator reaches a position where sound is currently being produced, record this as a capture, activate the food delivery mechanism, and begin a new random rest interval. If no capture occurs, once the prey has moved all the way through the random acoustic path begin the random interval between prey movements but do not deliver food.

By now many readers will have recognized that some of the recipes included in a chapter that is arbitrarily labeled "felines" could be useful in enriching the lives of many very different kinds of animals. The obverse is also true. Succeeding chapters include a number of recipes that could fruitfully be used with little modification for enriching the lives of animals in the genus *Felis*. For example, recipes for live fish feeding opportunities described for other groups of animals could be used with little or no modification to enrich the lives of many cats. In nature many cats are skilled hunters of fish and will relish consumption more when they are caught live than when they come in the form of fish meal or out of cans or freezers.

# 4

# Elephants

Elephants are truly wondrous, sensitive and intelligent animals, with a history of exploitation by humans. Their attributes may easily contribute to one's desire to do something to enrich the lives of captive elephants. It is no wonder that so many books have been written about these mammoth creatures, further enticing efforts on their behalf. If you choose to do enrichment work that involves direct contact with elephants, constant vigilance and safe procedures are required. Even the tamest and friendliest of elephants may cause unintentional harm because of their size and strength.

Every year a number of people are killed while working with elephants. Many more are maimed. The statistics concerning injury that are available from zoos and wildlife parks are doubtless underestimates, in large part because those working with elephants, including your author, have come to love them. We would not want an elephant to be punished, excessively confined, sold to some other place, or destroyed because of accidents that could have been avoided if only humans had exercised more caution. I have included in this chapter some personal accounts of dangerous or lethal events. My hope is that this will encourage readers who choose to do enrichment work on behalf of elephants to employ procedures that are as safe as possible.

## Our Earliest Work with Elephants

My first extensive opportunity to work with elephants was in the Portland Zoo beginning nearly 40 years ago. This zoo had a wonderful group of Asian Elephants including Thonglaw, the first herd sire for this species in captivity in the western hemisphere. Morgan Berry, the person responsible for donating this male elephant and others to the zoo, became a friend. Among the many things I learned about Morgan was his role in Thonglaw's first successful mating. Morgan had gotten behind and pushed and otherwise encouraged him to mount a female elephant when Thonglaw was at first reluctant. This heroic account was confirmed to me by eye witnesses.

This history-making push started the Portland Zoo staff on a long tradition of efforts to learn about breeding Asian elephants, and to help in their conservation. My own initial work with elephants would doubtless have been less rich in opportunity were it not for Morgan's kindness and contributions. His friendship and zeal were encouraging in my early hands-on work with these wonderful animals.

**Learning to Heed my own Words About Safety**

After seeing tragedies in other peoples' work with elephants, I thought that I would always do my best to use safe methods. But I confess to not always following my own advice. The following experience drove forcefully home to me that I was not immune to danger if my vigilance wavered.

Our research and veterinary team was helping a colleague from the University of Oregon Health Science Center with his systematic study of placental materials. As soon after each birth as it appeared safe, we collected the afterbirth materials from the floor of the elephant barn. It was in one of these efforts that I was the closest to being seriously hurt by an elephant in decades of work with them. This was an elephant that I loved, trusted, and had worked with for years. She shared quarters with the new mother and other female elephants and was standing just behind me, both of us quite excited by witnessing the birth of the elephant. I made the mistake of ignoring this gentle giant while I bent over to help with the placenta collection. One of my students screamed from her distant observation point and I sprang up to see a massive elephant trunk descending from on high. I heard a whooshing sound as the trunk barely missed my head and slapped loudly on the floor. Luckily, I had turned and stepped aside just enough to avoid being crushed.

This was an indelible lesson in teaching me to be more cautious and avoid potential harm in working with these wonderful animals. I also learned early in my efforts how important it is to exercise caution on behalf of others.

A typical behavioral trait is that for no understandable reason, there are some people that individual elephants seem to seriously dislike. When this is true, it has been my experience that it is best not to force the issue if you wish to maintain a positive relationship with the elephant by non-coercive means. That is, do not allow the chance for this human and this elephant to remain in direct contact!

When my son was young, he often spent time with me in my work with captive animals. One day I was working to develop a relationship with a young elephant in preparation for providing outdoor exercise opportunities in the zoo. Tim, who was about 6 years old, was riding on the elephant's back with my arm supporting him. The elephant was having great fun prancing around and keeping me running alongside. My effort in keeping up apparently entertained both the young elephant and my son.

An assistant came to the door to tell me that I had a visitor. It was a woman whom I had known since she was a child and who had worked with both domestic and exotic animals such as horses and felines. I lifted Tim down and invited her to join us and meet the young elephant thinking she might want to help in our efforts on behalf of this fine animal. The "gentle" elephant immediately charged her and pinned her to the wall. No vocal persuasion could gain her release, and I had to smack the elephant in a safe place on his side to get his attention momentarily so that she could escape.

I have absolutely no idea why this elephant did not like this kind young woman who so easily bonded with other animals. Repeated efforts to gentle him in her presence never succeeded. My advice arising from this and other similar experiences is rather simple. If you choose to do "hands on" enrichment with elephants do not stubbornly persist with an animal that does not like you, but turn over the work to others.

**Elephants at the Honolulu Zoo**

It was my good fortune to be teaching and researching one summer at the Honolulu Zoo when Indira Ghandi came to present a young elephant to the zoo as a gift "to the children of Hawaii." Along with the elephant came the mahout who had worked with this elephant for years. I learned much from this mahout that was useful and some other things that were quite unexpected. For example, the veterinarian working with us noticed that this elephant had cyclical diarrhea indicating that he had a parasitic infestation. With medication the parasite was eliminated and the diarrhea ceased. The mahout had become my friend and was kind and always ready to help and to share his thoughts about elephant well-being. A few weeks following the successful parasite treatment, he came to tell me that he feared the elephant was ill. When I inquired why, he told me through a translator "because elephants always relieve themselves by having diarrhea at predictable times and he no longer does." He earnestly believed that it was the *nature* of elephants to have diarrhea periodically on a scheduled basis. By eliminating endemic diseases zoos can certainly "improve on nature" in this respect.

There was a mandated period in which the new elephant had to remain quarantined from other elephants in the zoo. In anticipation of the arrival of this gift from India, the Corps of Engineers had voluntarily designed and built a lovely large stockade to serve as temporary housing. When there were no zoo visitors present the elephant keepers and the mahout would take the elephant into a large field for exercise and to teach useful commands to the keepers.

The mahout was great with the young elephant, able to gently control its behavior and to teach members of the zoo staff the commands to get the elephant to accomplish routine tasks (figure 4-2). This kind of training of animals along with those responsible for their daily care greatly serves to improve the lives of elephants in captivity.

Figure 4-2. Keeper Learning to Train Elephant

Providing competent human companions for non-dangerous animals and allowing the elephants to move into larger areas for exploration and exercise is enrichment that requires no apparatus. This intimate contact is also enriching for the humans involved and tends to develop a closer working relationship and companionship with the elephant.

**Contrasting Some Frequent Methods for Direct Contact with Elephants**

While conducting research with dolphins and enrichment work for a variety of animals at a theme park, I was fortunate in becoming friends with the trainers and the husbandry staff. The two trainers primarily involved in directing the work with elephants were of entirely different persuasions with respect to methodology. One persistently confided to me that he was desperately worried that the other would be hurt or killed by his elephants because he had not established a dominance relationship with them. Over the decades that have passed since we worked in the same venue, whenever I have met either of these two keepers it has been heartwarming for me to see that they both remain my friends even though I was not willing to take sides in their disputes.

The vast majority of highly respected animal trainers and zoo keepers that I have met believe, along with the first of my friends, that to work safely with elephants it is necessary to have a "private session" in which you physically establish your dominance. While some of us may not be cut out to involve ourselves in these efforts which inevitably involve some severe punishment, it is still possible to be sympathetic with this view if you have ever observed elephant behavior in wild or captive herds. Elephants are much more brutal in establishing dominance among their own group members than any humane elephant trainer or keeper is in establishing a working relationship with them.

Anyone observing my friend and the elephants which now treated him as dominant could see that he had a close bond with them and constantly evidenced positive concern for their well-being. He literally worried more about seeing that they had excellent health care than he did about his personal health concerns.

The other trainer was devoted to using only positive reinforcement methods except in case of emergency. However, as you read on you will see that using kind techniques and positive reinforcement may not always be safe for the human unless real caution is used. This trainer was a devotee of operant conditioning procedures and had remarkable success in training both elephants and camels to safely give rides to visitors in the theme park. He did this by giving rewards to the animals as they became progressively more adept at traveling designated pathways with humans aboard. I occasionally shared the joy apparent in his friendship with these elephants by helping with private sessions when the park was closed to visitors. We would walk them to the waterway that surrounded much of the park and take the elephants for a swim.

One day this trainer told me that he had recently obtained a very large female Asian elephant to be used in his work, and that he wanted me to be the first one to "ride on her ears." He said this in part because he had just done careful measurement to show that this was the tallest female Asian elephant in captivity. He exuberantly announced to me that she was taller than my favorite elephant from work at the Portland Zoo who held the previous record.

The elephant had trained well, knew the route, and would reliably walk calmly through it. After listening to me express all my reservations, this friend insisted, and I took the ride successfully with no problems along the way. The next day there was a giant helium-filled balloon raised at the perimeter of the park by the network preparing to telecast the world water-skiing championship. The elephant was so excited by this event that she tore out the entire side fence in that part of the park. While counting my blessings that it was not the previous day when I was astride her that this had happened, I spoke with the trainer about the good fortune that no one was riding the elephant when the excitement and destruction occurred.

He apologetically confessed to something that he should have shared *before* offering me the honor of the first ride. He had been able to obtain this fine elephant at little expense because she had a reputation as a rogue in the circus that had previously owned her. He was happy to have "rescued" her and to see that his positive reinforcement methods led to making her apparently tractable and happy.

Like all complex animals, elephants are individuals with varying personalities. At a later time, this same trainer was contractually invited by the Honolulu Zoo to do the initial training necessary so that a resident elephant could safely be walked through the zoo. The zoo director was anxious to give this elephant some relief from the hard substrate and excessively hot environment where she was on exhibit. I felt pretty secure that this would be a safe venture because the elephant had proven both interested in human interaction and careful to avoid injuries to humans in the many summers that I had worked with her. When I recommended this trainer to my friend the zoo director, I also told him about the incidents previously described.

When my friend's efforts in Honolulu were complete, the zoo keepers were leading the elephant around the zoo. She had been trained to help with cleanup activities by picking up trash and putting it in garbage cans along the walkways. Her "work" was aided by children who were allowed to follow at a safe distance and assist the elephant by picking up garbage and putting it in trash receptacles. The exercise and attention which she quite apparently enjoyed was good for the elephant, and the educational impact for kids was very clear. We observed participants later encouraging their parents to help pick up trash found on zoo pathways.

My recounting of pertinent events involving this kind trainer does not end on this happy note. He decided to get out of the business of show parks and become a zoo keeper. I went to visit him when he returned to work following a prolonged absence for recovery from injuries inflicted by a male elephant that attacked him. He told me that he knew this elephant was dangerous. Feeling that he had developed such a good working relationship with these animals, he got careless and unthinkingly did something forbidden by zoo policy. He entered an indoor area with the elephants to do some cleaning, when there were no other zoo staff members around. To make things worse, he had failed to leave an escape path for himself in case of emergency. He of course blamed himself rather than the elephant.

**Protected Contact with Elephants**

Significant progress has been made in techniques for training animals allowing health care and general husbandry to be accomplished more safely and humanely. A number of folks have become specialists in contracting to train staff of captive animal facilities in these methods. This is admirable work, and it is wonderful to see killer whales (*Orcinus orca*) "volunteer" their dorsal fins for blood drawing, apes extend their arms through cage openings for vaccination, and elephants who raise their feet so that necessary trimming and filing can be accomplished. Maintenance of healthy captive elephants requires this pedicure because their feet do not receive the kind of natural wearing down of hardened skin surfaces that occurs in their native habitats.

It is rewarding to view captive animals and husbandry staff interacting in positive ways, and observe care being provided without having to excessively restrain animals or use force. It remains critically important to be ever vigilant and mindful of potential danger. I once viewed a disturbing videotape in which one of the most prominent people employed for teaching protected contact methods is seen being attacked by an elephant that he briefly ignored. He was pointing at newly installed equipment and describing the method for its use in protected contact when the elephant's behavior caused his injury. Even when accomplishing work clearly intended to benefit animals and husbandry staff, constant attention to their behavior and whereabouts is necessary.

**Simple Enrichment Helps with Demonstrating Common Behavior for Elephants in Various Settings**

Many elephant populations have undergone generations of selection based on docility for domestication. It seems reasonable to expect that these animals might benefit from positive interactions with humans when conducted in a manner that is safe for both species. Among programs that I have seen work relatively well are demonstrations of typical work for domesticated animals, such as log moving. This provides exercise for the elephants and teaches zoo visitors about their tremendous strength and dexterity.

There are still an unfortunate number of captive conditions, including those in many zoos and entertainment venues, where elephants have no appropriate substrate and live on concrete or asphalt surfaces. Historically it was often deemed more important to be able to keep the exhibit clean of excrement than it was to consider the elephants' comfort and needs. Elephants in the wild often "bathe" themselves with dirt thrown over their backs with their trunks. One relatively easy thing to provide for them in zoos or theme parks is a part of the exhibit where this can be done. Although it may require regular raking of the dirt back into appropriate areas and a means to ensure that the tossed dirt will not clog drains, it is enrichment well worth considering. Providing exhibit areas with real or artificial rock edifices to use in scratching their backs has also proven popular with elephants. Visitors to zoos not only learn about elephant behavior where enhancements like this are provided for the elephants, but are often entranced by watching these wonderful giants shower themselves with dirt, traipse around on natural surfaces and occasionally scratch their own backs.

## Simple Recipes for Responsive Elephant Enrichment

Our own earliest extensive work with elephants focused both on some formal studies that gave credence to the well-circulated suggestion that elephants possess long-term memory (Markowitz, Schmidt et al 1975), and some efforts exclusively designed to provide enrichment for elephants (Markowitz 1982). For example, we were able to provide an opportunity for elephants to take an invigorating shower by yanking on a shower chain with their trunks (figure 4-1). This turned on the water jets surrounding the massive open doorway in their inside quarters. The kind owners of a car wash company had donated and installed this system at our request (and shameless begging). A brief recipe for a simpler and less expensive shower system is included later in this chapter.

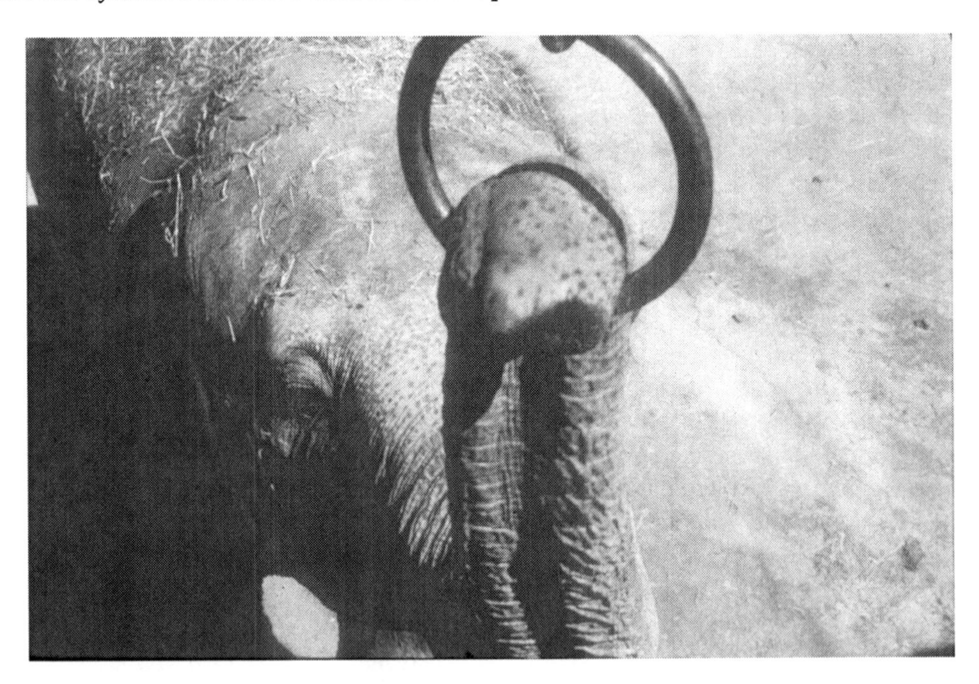

Figure 4-1. Elephant Pulls Ring to Turn on Shower

Before putting any apparatus within reach of elephants, it is necessary to remember their tremendous strength and remarkable ingenuity. That is why I recommend that, wherever possible, enrichment apparatus be out of reach of the elephants. Where that is not possible, it must be sturdily installed in a

manner that will prevent destruction by animals capable of pulling bolts out of concrete walls if there is some way to grab them. We watched with concern when an elephant reached a great distance outside her enclosure and turned over one of the heavy enclosure that enclosed both apparatus *and a researcher*, who was involved in our studies of memory abilities of elephants.
Fortunately neither researcher nor equipment was harmed. We breathed easier and got to tease and caution the researcher about moving the apparatus too close to the elephant's quarters.

## Water Showers

Many elephants will take advantage of opportunities to take water showers. This can be accomplished relatively inexpensively as long as there is a water line with sufficiently high pressure and a source of electricity in or near the elephant quarters. The first method will turn on a shower for a pre-selected time whenever the elephant actively "orders" it. The second will turn on the shower for as long as the elephant remains in the vicinity of the shower.

### Recipe 1. Inexpensive but Requires Local $H_2O$ and Electricity Sources

1. Mount an adjustable showerhead above the elephants reach or behind a wall over which the shower can spray.

2. Between the water supply and the shower head, mount a 110 volt AC solenoid valve wherever it is most convenient and meets safety and service requirements. Remember that the supply line as well as the valve must be out of the elephants' reach.

3. Select an appropriate manipulandum so that the solenoid valve can be turned on in a manner that is sturdy enough to be "elephant-proof" but still can be operated by these powerful animals. We have sometimes done this in the past by welding heavy steel pipes to form a bracket from which a chain with a steel ring on the end could be hung. Pulling the ring operated a switch that closed the circuit to operate a solenoid valve. Seeing the elephant pull a ring to turn on a shower is an attraction for visitors. However, this method involves some reasonably large expenditure for sturdy materials and for a good

craftsman with welding skills. A cheaper solution, if you have an appropriate heavy wall on which to mount it, is to use a large stainless steel plate that serves as part of a contact detector. The elephant can turn on the shower by touching the stainless plate with its trunk. Contact detectors do not require any movement of apparatus for a response to be registered to turn on the shower. The only thing that need be exposed to the elephant's strength is a plate imbedded in a wall in a manner which provides no edges for the elephant to pull on.

4. The device used to allow the animal to turn on the shower must have an output to operate an adjustable time delay relay. This relay should be selected with output capacity sufficient to safely provide energy to the solenoid valve.

In summary, this system allows the animal to make a simple response that turns on a shower via an adjustable timer. When the time expires, the shower goes off until the elephant turns it on again.

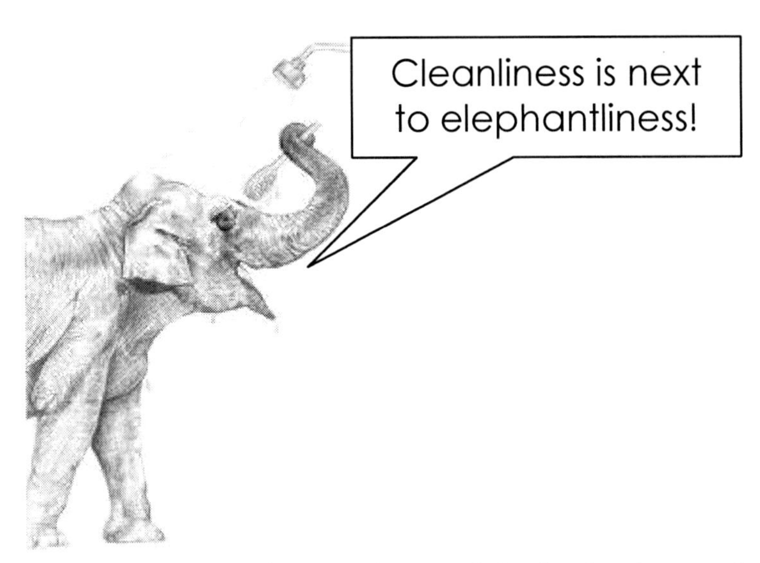


I prefer the recipe below for simple enrichment because it is simpler, less costly in construction, easier to make safe, and allows the shower to remain on only while an elephant is present to enjoy it.

**Recipe 2. Elephant Proximity Activates Shower**

1. Mount a driveway-type passive infrared motion detector (see Chapter 20) out of reach of the elephant.

2. Focus it on the area in which the elephant can receive a shower. Be certain to adjust the detector so that it will operate in daylight as well as darkness.

3. Install a solenoid valve and shower head to the water supply as described in Recipe 1, but this time activate the shower for the duration that the elephant is detected in the shower area. This is easily accomplished by connecting the output relay of the motion detector directly to the solenoid valve.

Besides the fact that this enrichment effort is low in cost, it requires only a modicum of technical expertise to assemble.

## Acoustic Enrichment for Elephants

An infrequently used and promising technique for elephant enrichment is the use of sound in enriching elephant lives. Some instances of sound as stimulation for captive elephants are summarized by Melo (1999). These include the playback of low frequency elephant calls of conspecifics for captive African elephants (Langbauer, Payne et al 1989), the use of an elephant-sized xylophone in the Phoenix Zoo (Schanberger 1991), access to an old fashioned organ grinder which the elephants wind with their trunks in the Berlin Zoo (Gibson, 1999), and trainer monitored use of a harmonica, tambourine, and electronic keyboard in the Houston Zoological Gardens. All of these techniques involve the need for human monitoring, and/or training. The inventive reader will undoubtedly recognize that it would not be difficult to use some recorded sounds that were reinforcing for elephants and provide them with the opportunity to turn on these sounds.

### Recipe 1. Elephant Turns On Sound By Moving to Focal Area

1. Place a motion detector so that it is focused on a limited area of the elephant habitat.

2. Adjust the timing controls on the motion detector for the desired length of sounds that will be played when an elephant enters the area.

3. Use a playback device of your choice (see chapter 20). Route the audio signal through contacts on the motion detector that close for the pre-selected time when elephant presence is detected. Attach the audio output from these contacts to speakers mounted out of the elephants' reach.

This is very simple and reasonably inexpensive enrichment that may be very effective assuming that you identify acoustic material that proves entertaining for the elephants. Use of a motion detector with an adjustable base will allow you to change the area where elephant presence turns on sounds, thus easily allowing husbandry staff to vary the area where sounds will be triggered.

**Recipe 2. Manipulation Alternative**

Identical to Recipe 1, except provide a rugged manipulandum of some sort that elephants can use to turn on the sounds and a timer to limit sound duration after responding.

## About Complex Response Requirements for Elephants

Because elephants are so clever and adept at performing tasks with their trunks, it may be tempting to devise elaborate schemes to allow them to solve various kinds of puzzles to "exercise their minds" in earning food treats. In an early study of elephant learning abilities, Les Squier found that elephants were so willing to work for sugar cubes that they would work at rates where their metabolic needs could not be met (Markowitz, 1982). Although sugar cubes are fairly easy to dispense, and a very powerful reinforcer that has served in research efforts, they are certainly not a recommended part of the elephant diet. For those who design enrichment that is constantly available for these animals, there are a great number of other things such as pieces of fruit or unshelled peanuts that can also serve as treats.

Elephants are not the most patient of animals when frustrated. They may either abandon excessively puzzling work, or in worse cases become aggressive towards other animals or tear up parts of their habitat. Consequently I recommend beginning with less difficult tasks and progressing slowly to more advanced problem solving tasks.

If elephants are kept in a large area where exercise can be encouraged, you might begin by distributing food treats in widely spaced places rather than building elaborate mechanisms to trigger release of food. This manual enrichment approach has the advantage that food locations can be varied from day to day, thus encouraging elephants to use more of their environment. Since their enormous size requires that they ingest large quantities of food, it is an easy (though weighty) matter to break the food supply into smaller portions and cart it to various locations.

To provide enrichment and encourage exercise throughout the day, you might then progress to some automatic equipment that does not require constant presence by husbandry staff. The two following recipes allow adjustment to fit the size and nature of the animal quarters.

**Complex Recipe 1. Acoustic Cue-Based Enrichment**

1. Install a number of waterproof speakers in safe widely separated places where they are audible to the elephants. The number of locations should be significantly greater than the number of elephants in the enclosure.

2. Test a variety of sounds to see which are most likely to attract elephants to approach the area where sound is heard. Use your observations to select a number of sounds that attract the elephants. This will allow you to make the enrichment more interesting by varying which sound is used each time an enrichment opportunity is provided.

3. Install a motion detector at each location to monitor elephant presence.

4. Program a computer to deliver the stored sounds at random times and locations for selectable durations.

5. Use inputs from the motion detectors to signal the computer when an elephant enters the area where the sound is playing. The computer program should allow flexibility in the number of movements that elephants must accomplish between sound locations in a timely manner to gain treats as rewards for exercise. This will allow you to progress slowly in teaching the elephants how to entertain themselves with sounds and be rewarded for healthful exercise.

6. Be sure that your computer program includes an adjustable limit on the time allowed for movement to each area before the sound disappears. When an elephant is detected in an area while a sound is playing, have the program turn off that sound and move it to another speaker location. When a pre-selected number of locations are entered during the presence of sound, the program must initiate delivery of treats. An inexpensive I/O board will allow you to easily connect inputs and outputs from motion detectors and the outputs for speaker sounds and feeder operation.

7. Consult chapter 20 in making decisions about the kind of food treat delivery device to employ. I recommend a 35mm film loop based dispenser and loading it with fruit, peanuts, and other nutritious treats.

In summary, this general recipe describes a method to encourage exercise and provide some acoustic stimulation for captive elephants. The use of a computer program allows you to vary the movement of sounds and the number of sounds that must be "investigated" by the elephants in a timely fashion in order to obtain food treats.

**Complex Recipe 2. Multiple Tasks For Elephants In A Large Captive Habitat**

1. Modify the above recipe by substituting some task for the elephant to accomplish each time it enters an area where sounds are being played. This recipe does not require the use of motion detectors, but still uses the computer program to limit the duration of sounds at each location.

2. Install some safely mounted equipment that allows the elephants to do some task at each area where sounds may be encountered. Each designer will have some favorite ideas of their own for elephant tasks, so I will just suggest a couple of simple ones to convey the concept:

Example 1. Install a sturdy telephone pole type log attached at one end on a steel axle in a manner allowing the log to rotate in a complete circle. Use a simple device such as a magnetic switch to detect when an elephant moves it through an entire rotation. Route the output of the switch in a protected manner to an input of your computer I/O board.

Example 2. Use a ground level steel plate with a sturdy auto spring beneath it and a magnetic switch or other device to detect when an elephant has stepped on it with sufficient force to depress the spring. Route the output of the detection device in a protected manner to an input of your computer I/O board.

The remaining steps are identical to those in Complex Recipe 1.

**Working On Behalf of Captive Elephants**

Looking back at this chapter, I realize that I may have left the impression that working with elephants was necessarily more of a dangerous task than a pleasure. Nothing can be farther from the truth once you develop a close relationship with these awesome creatures. My sweet wife would always insist that I take off my smelly clothes outside the house and head for the shower after working with elephants. I would happily do so while loudly recounting to her how much fun it had been to work with a particular elephant or two that day, and how responsive they had been. On days when our work helped to uncover a treatable medical problem, I was overjoyed and Krista teasingly lamented that I was "talking her ear off" while I was undressing and rushing to the shower. Being able to accomplish things that enrich the lives of elephants even in a simple manner is indeed a pleasure.

# 5

# CAMELS AND GIRAFFES

I have selected these two kinds of ungulates as the focus for a brief chapter because they are so often on display in zoos and wild animal parks and present some special problems for those who provide environmental enrichment. In general, the advisability of spreading out food rather than putting it in a single location is enriching for these animals in the same way that it is for most ungulates. It reduces aggression and allows more peaceful eating by less dominant animals.

## Camels

Watching desert movies in which heroes team up with camels to rescue maidens in distress may lead you to believe that they were born to a symbiotic relationship with us. This is far from the truth because until they undergo training for domestication, camels can be exceedingly unpleasant when approached by humans. Untamed camels quite willingly spit on, bite and viciously kick out at humans.

Because of their obstinate nature and reluctance to immediately respond to human commands, in some quarters, camels are reputed to be less than gifted intellectually. Grzimek (1972) has described how one widely heard expert on these "beasts" thought of them as cowardly, stupid, and apathetic among a wider range of pejorative monikers.

Grzimek also points out that the German term for dull-witted is "Kamel."

Camels are not stupid! Given even limited opportunity to invent a strategy to exploit their environment, camels show remarkable ingenuity. As part of our work with camels Vic Stevens chose to study the number of responses that camels were willing to emit at a time to gain brief access to alfalfa pellets. This was part of his effort to study "ratio strain" in a variety of vertebrates. The camel's required response involved pushing a panel that operated a microswitch.

We expected the camel to use its prominent proboscis (figure 5-1) to push the panel in much the way that some other mammals had been seen to do. However, this bright animal learned that he could vibrate his chin at high rates against the panel thus activating the switch and avoiding the effort required to make discrete responses. This was interesting but also worrisome because Adolph began to grind down his incisors with this method of responding. The apparatus was modified for his protection (Markowitz 1982). My belief that camels are brighter than they are sometimes given credit for has further been bolstered by the number of times that I have witnessed camels controlling their keepers or trainers rather than vice versa.

Figure 5-1. Camel Displays Dentition

It may be especially important in the breeding season, typically January through March for domesticated camels, to exercise caution in interacting with them, to spatially distribute food, and to find some diverting enrichment activity. During breeding season male camels become so increasingly aggressive that they have been known to seriously injure or kill humans, inflicting vicious bites with their powerful jaws and ample supply of teeth. In spite of this fact, camels have been domesticated in both Asia and Africa beginning as early as the fifth century BC (Grzimek, 1972).

The generalizations that I have briefly made about camels are just that. There are significant morphological differences between Bactrian and dromedary camels and also some important differences between subspecies. Before deciding upon a means of enrichment you will want to become knowledgeable about the species involved.

One Hump or Two???

Primarily to dispel the notion that these are totally ignorant critters, some readers might wish to choose specific enrichment tasks for camels to engage in and for the public to observe. The challenge here is to find something that the camels do willingly once you identify an effective source of reinforcement to encourage the activity. A variety of foods can be loaded in either of the devices suggested below, and different parts of the devices can be loaded with different foods. Wild camels feed on grass, herbs, and thin branches from bushes and trees, so your imagination can run wild.

The following are two general recipes that I believe would be effective and attractive to camels. The first is relatively inexpensive, requires no electrical power and, if the equipment is well constructed, should be largely maintenance free. The only drawback is that it requires volunteer or staff time to regularly load food throughout the day.

**Recipe 1. Simple And Moderate Cost**

Here are some steps to build a number of rugged heavy gauge "lazy-Susan" type apparatus, each with a dozen or more uniform size compartments that are large enough to hold a supply of alfalfa pellets or other preferred food attractive to camels. You will have to decide how many of these rotary devices to build based on available budget and space limitations. Providing multiple devices will reduce competition for their contents and allow visitors more opportunity to see that camels are capable of both working voluntarily and using some intellect in their efforts.

For this recipe you will need to have someone expert in the qualities of different plastics or other suitable materials and able to skillfully machine them. Materials that are not subject to warping with temperature change and that make good bearing surfaces will be needed for the base plates and the axles themselves. In our construction and maintenance of enrichment devices we have found Delrin™ to be especially useful where plastic surfaces move against one another. It is a good idea to make a set of templates out of cardboard and assemble them before beginning to make certain that the parts will align properly.

1. Use a two inch length of half inch Delrin rod for an axle.

2. Purchase a one inch thick twelve inch diameter disk of non warping plastic. Scribe a circle eleven inches in diameter on one side of the disk to use as a guide line.

3. Drill eighteen equally spaced 1½ inch diameter half inch deep holes with centers on the line you scribed in step 2. The recesses that you make in the disk in this step will serve as cups for treats for the

camels. Dead center on the disk, drill a hole that will fit the axle snuggly and allow the disk to rotate without binding.

4. Construct an enclosure for the rotary feeder made of at least one quarter inch thick plastic. Plastic thinner than that will not resist warping, weather effects, or destruction by the camels. Mount the enclosure in an area where the husbandry staff will have access from outside on the side opposite that where the camels work to deliver themselves treats. Recommended *inside dimensions* for the apparatus enclosure: height 4 inches; width 12 ½ inches; length (based on ¼ inch thick plastic construction) 11 inches. Before cutting the enclosure components or assembling the walls, read the remainder of the instructions and description of operation that follow below.

5. Decide on where you will be installing each of these devices and drill necessary holes for mounting. For example, if you are mounting to outdoor metal fencing, you may wish to use large diameter stainless steel J hooks to loop through the fence and tighten inside the enclosure with nuts; if you wish to mount to a concrete wall, you may wish to use concrete anchor shields and bolt into them from inside the enclosure.

4. There is an advantage to mounting the devices in an area on the perimeter where zoo staff has access from outside the enclosure and the camels do not. If you choose such a location, it will make loading the cups possible without removing the lid from the enclosure *if it is feasible* to make a hole in the side of the enclosure just large enough to load food into the cups and manually rotate the disk. This should be done only if you are certain the camels will not have access to this opening once the feeder is installed in their enclosure.

5. If necessary, modify the internal dimensions given in step 4. Different lengths and/or widths may be necessary if you have chosen to use plastic that is not ¼ inch thick for construction of the enclosure. Refer to figure 5-2 to assist you in deciding on any modifications.

6. Cut two openings on the piece that will be facing the camel. The first opening should be a centered vertical slot about 1½ inches wide and 2 inches high. The bottom of the slot should be about 1¼ inches above the floor of the enclosure. The second opening must be cut to size so that it allows the edge of the rotating disk to protrude outside the enclosure. This protruding part of the disk will allow the camel to rotate the disk with its muzzle. To avoid problems in construction, check with a template to make certain that there will be sufficient clearance so that contents of the cups will not touch any part of the side as the disk rotates through the aperture you have made.

7. Make one of the sides of the enclosure easy to remove by the keepers but not by the camels. This will facilitate cleaning of the device and removal for loading food if it is not feasible to provide an opening for this purpose.

8. In the base of the enclosure make a tight fitting hole for the bottom of the axle. Use a template to ensure that you locate this hole so that the disk will protrude enough for a camel to rotate the disk and that the disk will not touch any side of the enclosure. Securely attach the axle to the base of the enclosure by means of a bracket attached on the outside of the bottom of the enclosure.

9. Use a very thin nylon or Delrin washer to ensure that the disk will move smoothly on the base of the enclosure. Assemble the device and test for ease of operation.

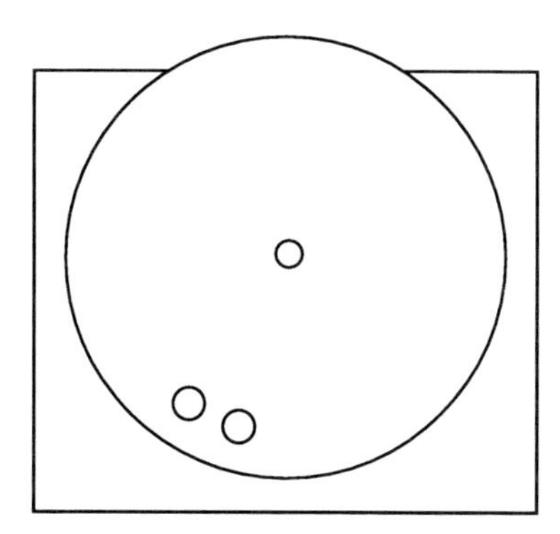

Figure 5-2. Rotary Feeder Partial Illustration: Edge of Disk Protrudes To Allow Camel To Rotate Treats Into Position for Gathering

**Recipe 2. Fairly Simple, But Requires Electrically Powered Components**

1. Install a series of waterproof lights at randomly spaced intervals around the enclosure perimeter, each centered above a sliding door that conceals a compartment that can be loaded with camel preferred food.

2. On each of the sliding doors install a locking mechanism that can be remotely controlled, such as the kind that is used for some entry doors in residences.

3. Use a computer program, or construct a simple electronic device to randomly turn on one of the lights and disengage the lock on the door beneath it.

The camels challenge is to open the door under the light with its snout. Users will undoubtedly envision additional ways to program use of this simple arrangement to add increasing challenges for the cleverest of camels. For those who are interested in testing the camels learning ability, time contingencies requiring that the camel go quickly to the door with the light on can easily be added,. For a greater challenge, the camel could be required to respond successively to some pattern of lights produced by your control mechanism in order to gain access to the food.

## Simple Enrichment for Giraffes

Giraffes, like camels, have 2 genera and only one species in each, but the eight subspecies of steppe-giraffes (*Giraffa camelopardalis*) differ both in appearance and native habitat. The Okapi (*Okapia okapi*) or forest giraffe bears small resemblance to members of the other genus because it has only a moderately long neck and many other distinctive features. Those planning enrichment are once again well-advised to become as familiar as possible with the species and/or subspecies of focus.

Watching giraffes awkwardly drink ground-level water, one can see that some effort is required for them to spread their front legs far apart or flex them to accomplish this task.

Fortunately the majority of zoos feed giraffes by elevating the food to a height where they can browse without bending over, yet there are a surprising number of institutions housing giraffes that do not. When one considers the naturally high blood pressure of giraffes, it does not make great sense to have them required to redundantly bring their great necks toward the ground in order to browse. So for those places that do not yet do so, a simple improvement that giraffes will enjoy is raising some branches or limbs with sufficient edible leaves for them to browse on comfortably.

Even in zoos with small budgets, this can be made relatively easy for husbandry staff by simply mounting a captive pulley and line on a pole or other convenient high anchor. Early in the day before the giraffes are released from night quarters, new browse can be attached to the line and raised up high.

Whenever possible enclosures that house multiple giraffes should have more than one such browse station in order to reduce competitive aggression for access to the choicest part of the browse. This is especially imperative in cases where new members are being introduced to an established group of giraffes.

Both in the wild and where possible in captivity, giraffes can be observed to pull high branches close with their long tongues and to strip the leaves from them. Whatever can be done to find suitable feeding materials that are fresh and ample and arranged at a convenient height will be rewarding for the giraffes.

For those wishing to add complexity for research purposes or for demonstrations, we have found that giraffes will also employ their tongues to operate contact detectors which are linked to providing food (fig 5-3). Another possible reason for wishing to require some active local response to initiate feeding is that it may help in separating animals during times that they are aggressive. Although fights with conspecifics are relatively infrequent in giraffes, when they do occur they may lead to serious injury. One reason for this is that giraffes are very prone to suffering from intestinal torsions if they are knocked down awkwardly.

Figure 5-3. Touch Apparatus High Above Giraffe Fence

Rather than arbitrarily selecting some recipe to detail for this kind of work, I will simply suggest one general idea. Devices that giraffes can lick or grasp with their capable tongues may be used to trigger the movement of browse to places accessible to the giraffes. One simple arrangement might involve moving the position of elevated browse from outside to inside their enclosure by means of a motor driven pole that swings the browse from location to location.

In closing, I return to the fact that the design of enrichment activities for these animals requires knowledge of their special adaptations. Giraffes and camels have been known to inflict serious injury on human handlers. Their front legs are powerful enough to ward off lions in the wild. When excited in captivity, they may kick out violently and cause injury to husbandry staff. It is important to plan ahead so that your enrichment does not introduce potential danger to either species.

# 6

# HIPPOS AND RHINOS

These are notoriously dangerous animals, and it is only in relatively recent times that much has been attempted in the way of enrichment on their behalf, or that exhibits have really been designed to meet their needs.

An important enrichment method for both species is one that now has been adopted in many zoos and wildlife parks. This will sound familiar if you have read previous chapters: Spread out the food and do not leave it in one concentrated area.

It is also advisable, when possible, to provide some places that are only accessible to smaller animals in the enclosure, so that they can eat peacefully and escape if necessary from attacks by larger aggressive individuals.

## Hippopotamuses

Although after a period of acclimation the pygmy hippopotamus (*Choeriopsis liberiensis*) often becomes more docile with respect to husbandry staff than does the large hippo (*Hippopotamus amphibus*), both can inflict serious damage when excited. There is much to be said for using enrichment apparatus that can withstand destruction and can be maintained and manipulated by keepers from outside the enclosure. That will be the concentration of this chapter.

Unless there is a very expansive environment for the hippopotamus, which is seldom the case in zoos, they are typically kept in smaller groups than those that occur in the wild. Where large areas may exist in wildlife parks, an important husbandry consideration is to provide places where they may mark territories by spreading feces. Such territorial marking and avoidance of these territories by others has been seen to reduce aggression in the wild. Therefore, overzealous cleaning of these areas is not recommended.

Battles in the natural habitat are frequently fought in the water and may lead to serious injuries and sometimes be lethal. It is important in designing places for hippos to enter the water that there be multiple narrower pathways rather than a single point of entry.

Both species of hippopotamus are significantly aquatic animals in the wild. Nothing is sadder than to see them with no real place to immerse themselves as they would regularly in their natural habitats. In addition to providing an area where they can enter water, and distributing food in multiple locations, it is enriching to include aquatic plants where possible. Feeding in aquatic locations is another important part of their natural behavior.

In those cases where really adequate bodies of water have been provided for hippos, I have sometimes heard complaints from zoo visitors about "empty exhibits" when a careful observer might have seen some ears wiggling just above the surface. The following three recipes all provide some partial solutions to the dilemma of increasing the time in which hippos may be seen involved in terrestrial foraging, while simultaneously adding some diversity to the hippos' captive life.

For all of the recipes presented here it is recommended that batteries be used to drive the feeder belts only when it is not feasible to bring wall current to the area. This is because the motors used to efficiently drive the belts consume enough power that batteries require frequent charging and replacement. Adding the chore of remembering to charge and routinely exchange batteries will make the equipment much less attractive to husbandry staff and may mean that it is only used intermittently.

**Recipe 1. Simple But May Be Costly**

Here is a simple method to allow a variety of appropriate food to be spread out over time as well as space. The major costs are for commercial-quality sanitizable conveyer belts, and to safely install electrical outlets for the equipment should there be none handy. An alternative for tight budgets is to make your own belt system if you have appropriate materials available.

1. Install at least three conveyer belts, either behind sturdy exhibit walls or in steel cases able to withstand the strength of hippos.

2. For each belt motor, attach a remote control relay with its normally open contacts used to supply energy. Remote controls capable of generating multiple signals are available at surprisingly low cost because of their widespread use in both industry and the home. Be certain that you choose one rated for outdoor use if you are going to expose the control to moist weather conditions.

3. If the feeder belt does not come with its own adjustable timing mechanism, and you do not want to hold down the button and watch for the food to emerge, it will be necessary to add a time delay relay to the apparatus. This will allow adjusting the length of feeder belt excursion for a single signal from the remote control. You will want to purchase a time delay relay that closes its contacts on a signal and opens them following the selected duration.

Loading the feeder belts can be done at a time when the hippos do not have access to the area. Husbandry staff can simply press the appropriate remote control button to advance the feeder belt sufficiently to drop one healthy portion of food for the hippos' consumption.

This will allow the delivery of food at random times in various locations, and will also be attractive when giving public tours or demonstrations (figure 6-1). Delivering food by this method may also serve as a source of diversion if hippos occasionally become hostile to other animals in their habitat.

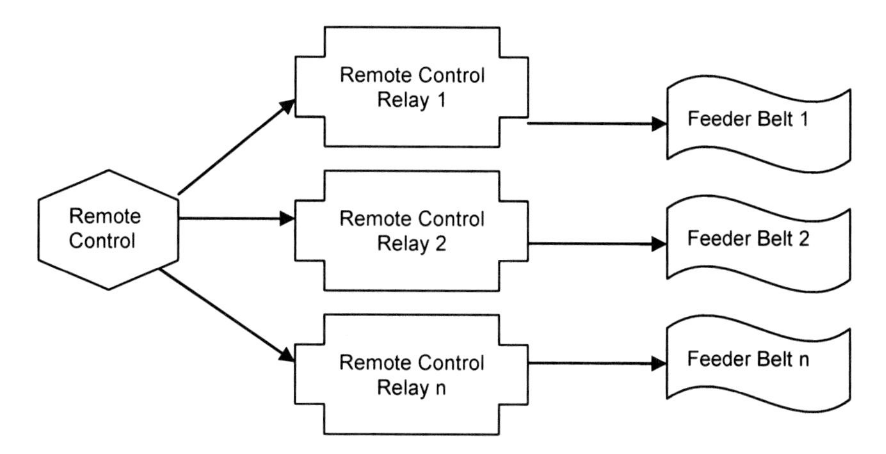


Figure 6-1. Block Diagram Simple Hippo Feeding Apparatus

Note that since a staple of the hippo diet is aquatic vegetation, one may choose to deliver some of the food in pond areas.

**Recipe 2. Automating Increased Probability of Responding When Visitors Are Present**

If you wish to have the probability of feeding in a particular location linked to the time when visitors are in a place to observe the feeding, this may be easily accomplished by modifying the recipe that has just been presented.

1. Instead of a hand-held remote control and associated relays, use motion detectors to trigger the closest visible feeding station when people are in the viewing area. There are two easy ways to accomplish this. Your choice may depend on whether the environment in which installation is made allows for direct wiring between the motion detector output and the feeder belt.

2A. If it is possible to run wire in appropriate weather proof conduit to the feeder belt motor, then all that one needs is a separate passive infrared detector (the "driveway" or "intruder" kind) for each station. Focus the device on the area where visitors are to be detected and adjust the range and sensitivity controls appropriately. Use the output to supply energy to the local feeder belt (figure 6-2). The duration of belt movement can be adjusted by using the controls on the motion detector that ordinarily adjust the time when a floodlight would be on in standard applications of such devices (chapter 20).

2B. For installations in places such as large naturalistic exhibits where it is expensive and impractical to run conduit between the detectors and the feeding stations, use the timed output of the motion detector to turn on a remote (wireless) relay switch. Deliver the energy for the conveyer belt across the normally open contacts of the remote control relay.

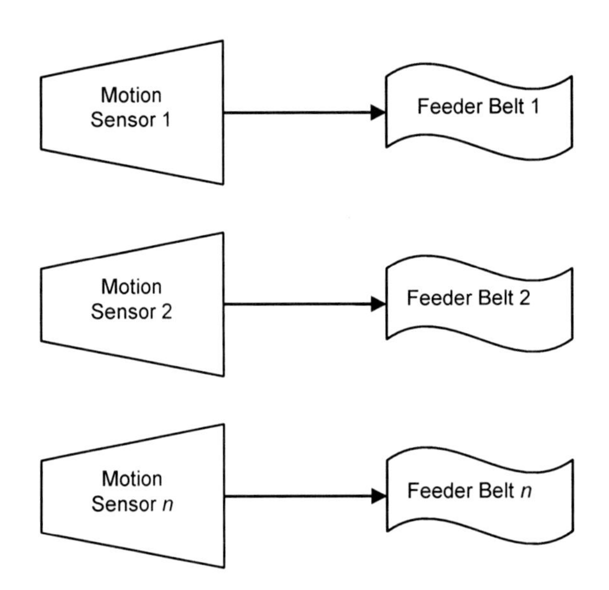


Figure 6-2. Motion Detectors Activate Hippo Feeders

**Recipe 3. More Expensive Signaled Food Availability**

Yet another modification of the above equipment may make it especially attractive for use in places such as theme parks with very large hippo exhibits. I originally designed such a system for a theme park where most visitors only viewed the hippo area while passengers on a monorail. The park administrators were concerned because the hippo was quite frequently submerged in a pool when the monorail passed, thus giving the impression that this was an uninhabited area.

This recipe requires two additions to the previous recipe's equipment. The first is a signal to let the hippos know where and when foraging for food will be successful, and the second is a motion detector that allows detection of the hippos near the feeder station. Where budgets allow, a little ingenuity will allow users of this recipe to disguise the feeder stations to look more naturalistic, and the foraging activity to appear more natural.

1. Use a digital chip (chapter 20) to record a different sound of your choice to be used at each of the automated feeding stations to attract the attention of the hippos. You might for example choose a

rustling sound at one, a bird sound at another and the sound of a terrestrial intruder at the third. Place each chip in a protected place in the appropriate feeder enclosure and attach it to an amplifier and waterproof speaker.

2. Choose either or both kinds of equipment presented in the first two recipes to trigger the potential foraging activity. If both are used, the advantage is that staff can trigger the equipment, or it can be activated when visitors are in an area that allows them a view of the places where the hippos are likely to forage.

3. To turn on the remote relay at the appropriate feeding station, use the motion detector or the hand held remote control, but this time have the relay turn on a timer apparatus designed as follows:

The output contacts of the remote controlled relay operate an adjustable time delay relay that is activated on signal and turns off after the selected duration. One set of normally open contacts on the time delay relay is used to turn on the sound. A second set is used to connect the output of the motion detector (focused on the area where the animal can forage for food) and the feeder belt timer. Thus, when the sound is on, if the hippo forages in the area this motion detector will activate the feeder belt time control (figure 6-3).

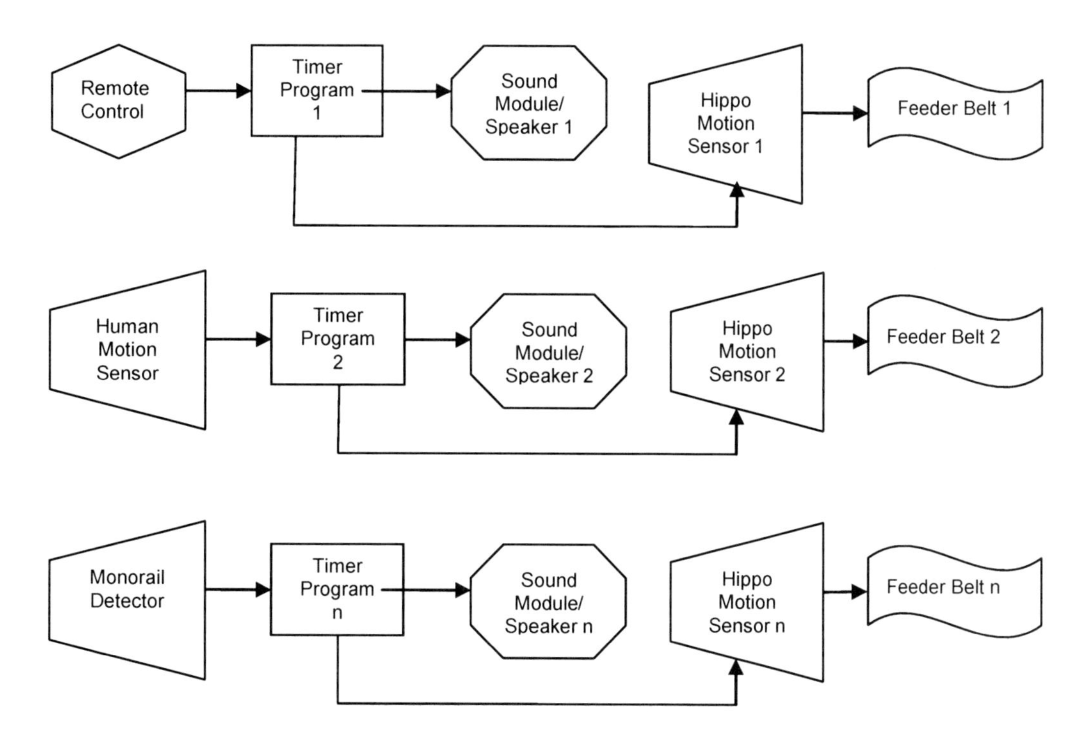


Figure 6-3. More Complex Hippo Apparatus Gives Hippos More Control of Feeding

For places that use monorails, the motion detector could detect the presence of a train, and/or the driver could use a remote control to turn on a local feeding signal. This might even lead the hippos to look forward with anticipation to the train arriving, instead of being annoyed by the disturbance.

In those places where hippos are kept singly or in very small numbers, it has been observed that young hippos are especially inclined to seek company of other compatible species when possible. There are even documented cases where they have solicited or welcomed interactions with animals ranging from dogs to elephants (Frădrich & Lang 1972). This suggests that an important form of enrichment for lonely hippos may be the presence of other animals where this can be safely managed.

When they are still under their mother's protection, young hippos remain very close and are kept on the side away from adult males and other perceived dangers by mom. She will butt them with her head if the youngsters stray too far from her. This companionship between mother and young is a critical part of the healthy development of hippos. It is unquestionably important to see that hippos be provided appropriate space and opportunity for this nurturing in captive circumstances.

## Rhinoceros

Although two of the species of contemporary rhinoceros live on the African continent and three in Asia, they generally resemble one another and regardless of differences in size, coloration, number of horns and other attributes, behavior across the rhinoceros species remains relatively consistent. All rhinos are herbivores, some being almost exclusively grass eaters while other forest-dwelling species may also consume browse. Today they all live in remote areas, in considerable part because of human exploitation of their natural habitat and slaughter of rhinos.

Readers will undoubtedly be aware of the fact that rhinos have long been coveted for their horns because of their reputed qualities when ground up and taken as "medicine." This continues to occur despite the fact that controlled studies have shown no measurable effect of such ingestion in humans. Today smugglers can get tens of thousands of dollars for a good-sized rhino horn, although I would prefer to see them get tens of years in prison for profiting from the killing of a rhino.

In the wild, rhinos can be observed to build dung heaps marking entrances to forest pathways. In the cases where there are truly forested semi-captive homes for rhinos, it may be an important form of enrichment to provide opportunity for such behavior. These marked pathways are often found near the rhinos' wallows or other bathing and watering places.

Rhinos are second in size to elephants among living terrestrial animals. Until recently, the black rhinos (*Diceros bicornis*) were by far the most numerous remaining in the wild, but today they are outnumbered by white rhinos (*Ceratotherium simum*) and Indian rhinos (*Rhinoceros unicornis*).

Although they can swim and do bathe often in the wild, rhinos are not the largely aquatic animals that hippos are in nature. Their typical day in the wild does not include the marking and defending of territories characteristic of so many other ungulates. Encounters with conspecifics are most often peaceful in nature. Rhinos may be seen cavorting with each other doing mock chasing while hissing and snorting around waterholes at night. Only afternoon hours are typically spent grazing. In areas where they are protected from predation and other human disturbance, rhinos spend much of their time lazing around in the shade of trees or in wallows.

Since rhinos have acute hearing ability, the third recipe given for hippos in this chapter should also be a naturally attractive form of active enrichment for rhinos. Any of the three hippo enrichment recipes might also be employed effectively for rhinos. Therefore the remainder of this chapter will be devoted to the description of various means of passive enrichment that have been employed by husbandry staff in zoos and other facilities that house rhinos. These are primarily gleaned from reports in *Animal Keepers' Forum* by Kayla Grams and Gretchen Ziegler (1995) and by G. Ziegler and Jan Roletto (2000) and from personally visiting with husbandry personnel around the world.

### Successfully Employed Passive Enrichment for Rhinos

Obviously herbivores do not have prey, but the near-sighted rhinos do somewhat indiscriminately attack any moving object that startles them, including humans and other animals. Consequently, a number of zoos have chosen to provide some rugged large objects for captive rhinos to attack and batter around. Where it is possible to do so safely, I recommend that there be some means for husbandry staff to occasionally pull attached chains or otherwise manipulate these target objects sufficiently to draw attention to the objects. I have seen rhinos apparently enjoying themselves by attacking inanimate objects, and doing this especially often when the objects are moved occasionally by husbandry staff.

In the Copenhagen Zoo a water pool and a mud pool have been provided, and rhinos may be seen to enjoy rolling themselves in puddles of mud and bathing on hot days. Rubbing posts are also included in their environment to promote stimulation of natural skin growth and peeling off dry skin.

This zoo also has an interesting program in which they exchange rhino dung with other zoos. The new scents have been seen to stimulate sexual activity and increase what appears to be marking behavior. The general husbandry procedure involves occasionally splitting up the rhino herd, since increased sexual behavior is seen when they are rejoined. Finally, browse branches are hung in high places to stimulate natural foraging behavior for forest rhinos.

The San Diego Zoo and many other facilities world wide place frequent browse that is distributed on the ground or hung on tree stumps. At the San Diego Zoo large movable stumps are spread throughout the enclosure. Along with many other zoos, San Diego employs large plastic drums for males to "aggress" against. Apples are put in the pool and consumed by rhinos. Fresh dirt is brought in as needed, and mud puddles are developed. This zoo uses elephant dung as well as dung from other groups of rhinos. When this technique is employed, males will defecate on the dung and stomp it into the ground!

"Snowmen" have also been built for various species of rhino in zoos. Components such as apples for eyes, carrots for noses, and bananas for mouths are used. Typical responses of the rhinos include exploring the strange "creature," consuming some or all of the edible components, and then battering and stomping what remains.

Finally, some zoos have employed "tests of strength" in which carefully secured objects can be moved with great effort by the rhino, and male rhinos seem to welcome this challenge.

In conclusion, there is much that can be done to enrich the lives of captive rhinos without enormous outlay of money. It is a shame that not all facilities housing rhinos provide some stimulation for their natural behavior. There continue to be far too many places that prize clean appearances and naturalistic "backdrops" more than opportunities to enhance the lives of resident rhinos.

# 7

# BEARS

Bears are an amazingly diverse set of creatures. Creative animal lovers have a seemingly endless array of opportunities to enrich their lives in circumstances where they are kept in captivity. Although they may all have omnivore chow as a staple in their captive diets, they are much more enticed by some of their favored natural prey. This makes it possible to use a variety of living and dead food sources to promote healthful species-appropriate activities. When dealing with those bears that especially favor large mammalian prey, such as polar bears who feed largely on ringed seals (*Phoca hispida*) when they can find them, no zoo feeds the preferred natural prey. It is both impractical (and seen by most people to be inhumane) to design even the largest captive mixed species environment allowing bears to capture and consume seals. On the other hand, a majority of visitors are tolerant of providing opportunity to feed on living insects and, in the right circumstances, on a variety of fish. Both of these kinds of prey are important components of the diets of many bears in the wild.

It is certainly possible to introduce a wide variety of vegetation as a form of enrichment for bears. But many species that lean toward the carnivorous end of this group of omnivores only become less carnivorous by consuming vegetation and eggs when prey is not readily available. This chapter will not explore the entire range of specialization in bears, but will instead concentrate on a few species for which we have had the most opportunities to design captive enrichment. Recipes provided will range from very simple and inexpensive enhancements to quite complex scenarios. Thoughtful readers with inventive minds will find it an exciting challenge to study the natural history of some of the other *Ursidae* and to invent ways to improve their captive existence.

## Polar Bears

For many years these most dangerous of predators were listed in their own genus, *Thalarctos,* but today they are classified in the genus *Ursus* along with other bears, while maintaining the species name *maritimus,* and as marine mammals because of their significantly aquatic nature. Unfortunately, most zoo visitors have historically been introduced to polar bears who reside primarily on concrete surfaces. Often there have been tiny, if any, swimming places for polar bears. Some thoughtful keepers have found that providing even the simplest of enhancements, such as strong empty kegs for them to play with, enrich polar bear behavior. Such additions are certainly an improvement on totally barren grottos, but much more can be done to improve the captive circumstance.

In much of their range, polar bears are solitary nomads except at mating time and when mothers are with young. This has been altered considerably by the presence of humans in areas such as Churchill in Manitoba, Canada. Annual arrival of polar bears has become a spectacle as they raid the garbage and search for whatever else they can consume as a function of human activity. Entrepreneurs now solicit visitors to ride in armored trucks to observe the large number of bears and to marvel at their power and aggressiveness in finding food. In zoos, visitors are inclined to favor exhibits that house many animals rather than just one or two representatives of a species, perhaps partly because of the faulty notion that socialization is always enriching at all times for all species.

It is always wise to test whether animals show sustained interest in what you propose to provide as active enrichment. This is true even if it seems intuitively obvious that they should love it. Relying on your empathetic ability and/or knowledge of the natural history of the species will not be sufficient to make intelligent choices in a number of cases that may surprise you. This is especially true for animals like polar bears that tend to be both highly individualistic and idiosyncratic in their behavior. Individuals that have lived most or all of their lives in captivity may fail to show the same preferences as their wild counterparts. Pre-testing your ideas in simple, non-costly ways may save much money and time.

Some examples of our own false judgments may be useful in bringing home this point. It is especially easy to make such misjudgments when you look at polar bears that are so clearly "outfitted" for cold weather and see them in a barren exhibit baking in the sun. All of us in our Portland Zoo enrichment group thought that it would undoubtedly be attractive for polar bears to have snow available on hot days. We made some sketches and apparatus diagrams while thinking about ways to allow the bears to order batches of snow whenever they wished to do so. I spent some time and money lobbying a company that made machines capable of producing snow-like ice chips to donate equipment to enrich the lives of the bears in our care. Fortunately, before squandering any more design or solicitation time in this venture we finally thought to make an evaluation of how the bears would view this "treat." Many buckets of shaved ice of the proper consistency were toted by hand to the polar bear grotto and tossed in for the residents' pleasure. As it turned out, their total response on this very warm summer day was to approach and sniff the snow pile and then totally ignore it while it melted!

Years later, a very kind keeper in the San Francisco Zoo came to me to ask if we could make automated showers available for the bears in his area of responsibility. He had noticed that some bears liked to bathe and play in the showers that he would manually provide by means of a hose from above the back of their grotto. My initial proposal was to begin with polar bears, since it seemed that they would especially enjoy the opportunity to turn on a shower or a mist on warm days. Fortunately, armed with the hard gained knowledge from the past, we invested no time or money in the system to do this without first systematically testing to see if the animals were attracted to such an "apparent" treat. It turned out that when I turned on the shower hose these polar bears actively responded to it.....by always moving as far away as possible! In contrast, one of the bears in the next grotto clearly enjoyed the shower and would rush to get under it wherever it was moved.

**Simple Recipe 1 for Bears**

1. Test to see if the bears like the opportunity to receive showers or mists. Be persistent and do not just try turning it on or off a few times, since it may take time for the bears to explore the shower opportunity and to habituate to your presence above their captive home. Stop at this step if the bears avoid the shower. Search for other potential means of enrichment (or more receptive bears).

2. Install a shower head in a logical place that will allow husbandry staff to adjust it from a fine to an intense spray depending on the bears' apparent preference.

3. Install a solenoid valve to allow turning the water on and off in the water pipe serving the shower.

4. Use a passive infrared motion detector (Chapter 20) for the device that triggers the shower onset.

5. Focus the motion detector on a part of the bears' environment that is not routinely traversed and is near the area where the shower will be delivered. Note that, as described in other chapters, all of this can be inexpensively accomplished. Choose a good quality water-resistant motion detector that has controls for adjusting the output duration and modify it so that instead of controlling flood lights, the output controls the solenoid used to turn the shower on and off.

The bear can now order a shower when it wishes by simply going to the part of its home where animal presence will automatically turn on the adjacent shower for a measured period of time.

## Sun Bears

One of my enrichment designs for sun bears (*Helarctos malayanus*) was a complex computer-controlled system for a new large exhibit at the Oakland Zoo. It worked well on installation and Anita Gavazzi, a brilliant graduate student of mine, wrote a wonderful program that provided easy adjustment of all of the parameters of each enrichment component. The program also collected detailed data concerning use of various enrichment opportunities. This complex system provided diverse enrichment. For example, there were opportunities for the bears to forage where they heard acoustic bees buzzing and to pursue the bee sounds until they arrived at the top of a tall tree stump. Successful pursuit led to honey (or other less viscous syrup) automatically being pumped into a reservoir atop the tree stump. Bears could also enter a small pool where their presence was detected and triggered release of fish into the pool. The presence of visitors was also automatically detected and increased the probability that the bears' enrichment availability signals, such as the buzzing of bees, would occur when humans were in the kiosk overlooking the exhibit. Zoo personnel giving tours were provided a wireless remote control which could also be used to initiate enrichment activity.

Unfortunately, the only zoo staff person with electrical experience adequate to service the equipment on his own left the zoo shortly after the exhibit was done. This was followed by an accidental flooding of the entire area in which the control computer was supposed to be safely protected. I present this somewhat detailed account as another example of why it is unwise to install complex enrichment equipment without the guarantee of permanent personnel on staff to see to inevitable maintenance needs. In retrospect, I should have designed individual active enrichment components for this exhibit. That way all aspects of the enrichment would have been unlikely to be disabled at one time, and equipment maintenance could have been simplified.

### Simple Recipe for Sun Bear Enrichment

1. Install a passive infrared motion detector in a region of entry to the bears' pond or pool. Modify the detector as described in previous recipes.

2. Use the timed output of the detector to deliver live fish into the water by triggering a delivery device which is hidden in a sturdy enclosure and protected from destruction by the bears. An alternative, where environmentally feasible, is to place the feeder outside the bears' enclosure and deliver the fish through PVC tubing to part of the body of water adjacent to the perimeter of the exhibit. This alternative allows loading of fish at any time rather than just when the bears do not have access to the exhibit.

**Recipe for More Complex Foraging Enrichment for Sun Bears**

1. Select some appropriate food treat for the bears, such as honey.

2. Program a digital recording device with a sound that might logically lead the bears on the path to finding food. For example, a playback digital chip can be used to present the sound of bees buzzing.

3. Install weatherproof speakers in a manner that protects them from destruction by the bears. The speakers should be arranged so that the sounds will occur in logical places for the bears to forage in hunting for some preferred food. In this sample recipe, hollowed-out stout tree stumps or artificial stumps house the speakers and provide a place where the bears may forage in hunting for honey. Deliver the sounds to the speakers either through protected conduit from the sound source or by broadcasting the signals by short range radio equipment to inexpensive radio receivers attached to the speakers.

4. Install devices to detect the bears foraging activity in the areas where sounds are produced. Rugged contact detectors or narrowly focused passive infrared motion detectors are logical choices. Either run wires from the foraging detection device through protected conduit to the apparatus control device, or use a wireless remote control to signal the control device when bears are detected foraging (figure 7-1).

5. Install a delivery device for the food treat and connect the input of this device to the appropriate output of the control device. We used a commercial liquid syrup delivery device to squirt honey into a stainless steel receptacle located in the top of a taller stump. Less expensive and easier to maintain solid food dispensing methods might be more desirable for many applications of this general design.

6. A simple computer with appropriate input/output connector, placed in a protected area, may be the least expensive and most versatile alternative for interfacing the equipment and producing appropriate sequential activities. It is also possible to use an industrial controller or to build hardwired equipment from inexpensive integrated circuits and solid state relays. However, these latter approaches require design and fabrication time and are recommended only for those with electronic experience. In any case, the equipment is programmed as follows:

A. At random or pseudo-random times, the sound is turned on at one of the foraging locations. The foraging detector at that location is simultaneously connected so that an output will be recorded if a bear forages while the sound is still on. If the bear does not forage in the pre-selected duration for sound, a new random interval between opportunities begins.

B. If foraging is detected at the area where the sound is being produced, the sound moves to another foraging location, and an identical hunt to that described above is produced. The number of foraging stations installed and the number utilized in any "hunting" sequence is arbitrary and depends mostly on the available budget and space.

C. If bears forage at each of the locations in the current hunt during the time the sound is present, the food treat is delivered at the final foraging area. The least costly design requires that a single location always be programmed to be the last of the required foraging locations. Only one device is then required to dispense rewards for foraging. If budget allows it would be more naturalistic to use a dispenser at each location in order that success in foraging is not always rewarded in the same place.

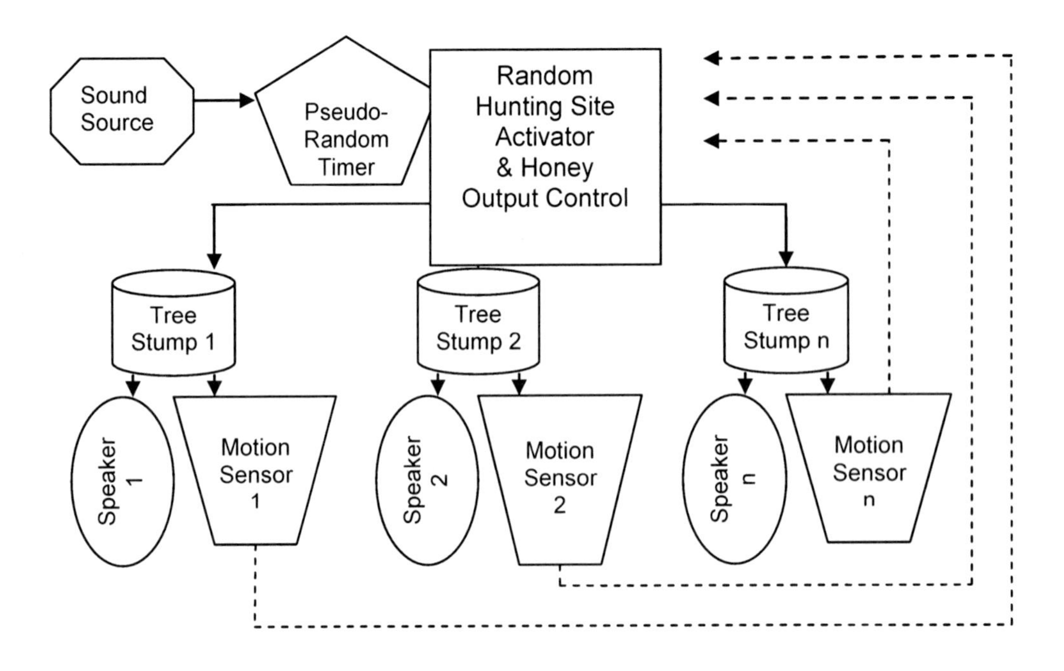


Figure 7-1 Honey Hunting Enrichment Components

## Problem Oriented Enrichment for Bears

Here are a couple of ideas derived from research and enrichment initiated because of husbandry and veterinary staff requests for possible solutions to ongoing problems. In both cases I have included brief anecdotal outcomes of the work on which the recipes are based. These may be useful should you decide to adopt one of these kinds of problem-oriented solutions. More detailed and illustrated descriptions of this early work can be found in *Behavioral Enrichment in the Zoo* (Markowitz, 1982). Both of these solutions had the added advantage that they produced activity in the bears which helped in their regulation of body weight and increased vigor (Schmidt and Markowitz, 1977).

### Recipe 1. Responsive Enrichment Aimed At Reducing Aggression

Many zoos still maintain bears in very limited grottos and attempt to display two or more bears with insufficient room for them to gain privacy or escape from aggression. The system described in this recipe is based on some work done with polar bears in the Portland Zoo. The male and female bear were frequently put on display together in the grotto and occasionally got into very nasty scraps that included wrestling and biting. The onset of fighting could often be predicted when the bears began to make nasty sounds at each other.

We first designed a system intended to see if reducing the percentage of hostile vocalizations by providing rewards for non-hostile utterances from the bears might reduce aggression. This was done by designing an acoustic filter to eliminate the major aggressive sounds as effective responses for a vocally triggered system that allowed the bears to "order" food treats to be automatically delivered. Although the enrichment effort ultimately worked remarkably well, I clearly made a blunder in first designing equipment with only one fish delivery site to deliver treats. This did not reduce aggression, because the clever female learned to stand near this delivery site and poach the food when the male vocalized in the area where "orders" were placed. Unlike your author, you will all undoubtedly be clever enough to realize that such a solution requires the simultaneous delivery of food in more than one location in the habitat.

We made the needed adjustment by adding a second treat delivery system that catapulted chunks of fish into or near the exhibit's pond. Using two fish delivery places significantly reduced aggression between the bears, and the results were interesting to observe. The only complaints came from some zoo visitors who could not get the bears to beg them for food as they had in the past. The bears were too engaged in ordering healthy treats and pursuing them as they flew into the grotto. While the PR folks found diplomatic means to explain why this was better for the bears, our research and veterinary crew privately rejoiced at the disenfranchisement of zoo visitors from their "historical right" to make beggars of bears.

Ultimately the male polar bear favored diving into the pool (figure 7-2) when fish flew in. He left the majority of ordering of treats to the female. She obtained fish in a location more proximal to the area where vocalizations triggered the delivery of some fish nearby, and simultaneously catapulted some into the pool.

Figure 7-2. Polar Bear Pursues Fish "Ordered by Spouse"

Here is a general suggested procedure for producing this kind of solution aimed at reducing aggression:

1. Make careful systematic observations of the behavior of the bears for a period of at least several weeks prior to designing any apparatus. Specifically aim your data collection method at identifying the most frequent behavior that predicts when aggression is about to occur.

2. Brainstorm with coworkers and experts with knowledge about the species and/or the focal captive bears. Concentrate observer effort and data collection on identifying behavior that is *never* observed as a precursor for physical aggression.

3. Before designing and constructing automatic responsive enrichment equipment, try manual approaches to assess the probability that your intended solution will work. This might, for example, be done by waiting for the bears to spontaneously produce the behavior that you wish to encourage and see if rewarding this behavior reduces the probability of aggression. Significant time spent in this labor-intensive effort will be more than compensated for since it will save you from building potentially useless equipment.

4. Having identified a behavior which, when rewarded, reduces the potential for aggression, design a method by which this behavior can be used by the bears to deliver something that is consistently attractive to them. One example would be a filtered microphone system such as we used in Portland to allow only non-aggressive vocalizations to trigger fish delivery. For some bears, you might use a device responsive to their gently rubbing against some surface in the exhibit, if that is a behavior observed never to be associated with aggression.

5. Install as many treat delivery sites as necessary so that bears will not aggress against each other to obtain the treats.

6. Design control equipment for the enrichment that will allow husbandry staff to adjust how frequently opportunity for engaging in the behavior will be rewarding for the bears. Include a means of signaling the bears so that they will know when their efforts may be rewarded. We have found that acoustic

signals are particularly useful in attracting the bears' attention because the bears do not need to be oriented in any particular direction to hear that a chance is at hand.

7. In summary, your equipment when finally produced will simply give the bears incentive to emit some particular behavior likely to reduce aggression and deliver them rewards for this behavior. A very general block diagram for such a system is shown below (fig 7-3), but inventive designers will likely find ways to improve on and embellish this simple suggestion.

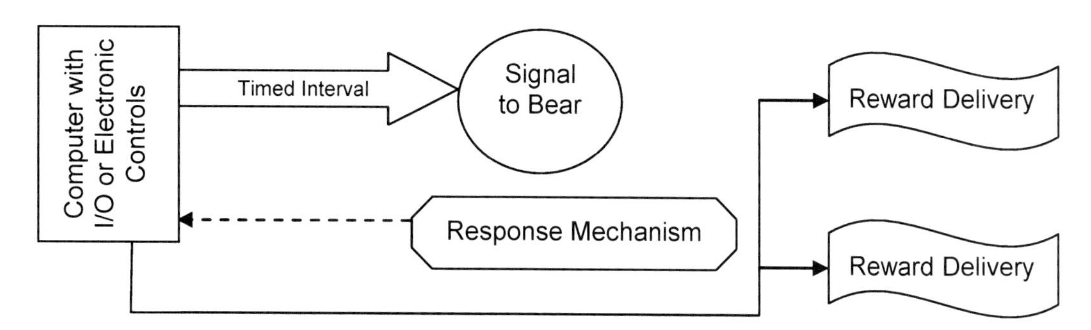


Figure 7-3. General Arrangement of Devices for Bears to Order Food Upon Signal of Availability

**Recipe 2. Responsive Enrichment Aimed At Interfacing The Public With The Activity Of The Animals In Safe Ways**

A progressive zoo director in Honolulu, working within the constraints of a very limited budget provided by the city, asked if we could help with a significant problem that he had. The difficulty arose from an improvement he had made in one of the bear exhibits. For many years the lone bear living there had been verbally encouraged by zoo visitors to beg them to throw in scraps of food that they brought from home. Not infrequently, the food that was thrown was selected because it was no longer judged to be suitable for human consumption but "okay for animals." The bear had settled into a daily routine of remaining sedentary in or near a pool in his exhibit. When visitors solicited him to beg, he would grab his hind paws with his forepaws and open his mouth. The visitors somehow found this behavior attractive, and the bear discovered that it would routinely bring him gifts.

The zoo director wished to eliminate, as much as possible, the unseemly begging behavior and the throwing of rotten food to the bears. So he surrounded the visitor viewing area with vegetation that prevented them from gaining the ballistic angle necessary to successfully throw food to the bear. This simultaneously improved the appearance of the exhibit and led to the reduced consumption of spoiled food by the bear. It also brought a torrent of local complaints to the zoo office and in "letters to the editor" published in Honolulu newspapers.

For example, a young woman complained that the zoo had now eliminated one of the true joys that had brought her to the zoo 'all the time she was growing up in Honolulu'....feeding the bear "treats" that she brought from home. Readers may well respond to this brief summary by thinking that what was needed was graphics and public education. However, these solutions are notoriously fraught with difficulties and were not successful in this particular case. One difficulty is that graphics are seldom read by visitors and almost never attended to carefully by those who come exclusively to be entertained by animals rather than to learn about them. Another problem is that politicians, who frequently control zoo purse strings directly or indirectly, are most often not happy with educational solutions that attempt to "force" the public to give up pleasures that they have historically enjoyed.

Ron Fial and I designed an apparatus that brought popular approval by providing a means for public interaction. Zoo maintenance shop personnel helped us with the installation and construction of some parts of the equipment, and students observed behavior of both the bear and the public once the equipment was available for use.

This enrichment apparatus included a rugged button which zoo visitors could press to catapult food into the bear's grotto. The food was attractive to the bear, and was carefully selected by zoo staff to be nutritious and fresh. An important feature of the catapult apparatus was that it delivered food from a protected food belt to random areas of the grotto. This was intended to encourage more activity by the bear who many staff members thought was "too old and arthritic" to move around much. Both visitor communications to the zoo office and media attention to this exhibit were now very positive.

My students and I observed two difficulties when the enrichment device was employed. It was so popular that on busy days rude adults and teenagers would sometimes be seen edging aside children who were queued in the line to push the transluminated button which indicated the opportunity to catapult food when it was lit. The other difficulty was that the bear did not become as active as we had hoped he would. He was too clever for us. Instead of actively pursuing each morsel that flew into his grotto, the bear simply observed until a number of food items were delivered. He then just got up occasionally from his resting position to make a food gathering excursion before returning to his sedentary mode.

We discovered no easy solution for the impolite behavior of humans toward each other. But in a later summer effort we did install additions to the apparatus to encourage more activity by the bear. While we wished there was budget to produce more naturalistic equipment tough enough to resist destruction by the bears, very few dollars were available for the work. So we had the zoo maintenance crew install long pipes that served as manipulanda for the bear to move by standing on hind legs and using forepaws to hit them. These pipes descended from opposite sides of the rear of the grotto and were mounted from the top of the back wall in a manner that operated a switch when the pipes were moved by the bear's paw.

In final form, the enrichment effort encouraged the bear to move between the manipulanda and operate them sequentially. This turned on the public apparatus and signaled visitors that they could catapult food to the bear. The bear indeed proved to be much more mobile and less apparently arthritic than some had feared he was, even capturing a kit fox that also lived in the grotto (Markowitz, 1982).

We have consistently found that inventing means for the public to participate in activating enrichment opportunities for animals has salutary outcomes. One is that it greatly increases visitor time spent observing the animals at each of these enriched exhibits. Another is that it reduces public attempts to gain the animals' attention by such actions as yelling at them, pounding on exhibit windows or walls, or mocking the animals.

Here is a very general recipe for interfacing the public with enrichment activity for bears:

1. Spend ample time observing the resident bears and public response to them in their captive environment. Concentrate your observations on identifying things that might productively enhance the animals' lives and reduce any possible deleterious behavior on the part of the public.

2. The most effective encouragement for healthful activity for bears is food treats, so select an appropriate food delivery mechanism for the food that you wish to deliver.

3. Test to make certain that the bears are willing to expend effort to obtain this food treat. For example you might try using an appropriate sound to attract bears to forage in some part of the exhibit and then deliver them food treats when they show foraging behavior.

4. Based on your observations of bears in this captive habitat, decide whether it is best to have the bears initiate the activity cycle or to allow visitors the chance to initiate the foraging opportunity. I tend to prefer the approach shown in figure 7-4 because the bears have control over when they will have an opportunity to actively forage for food. The public has the opportunity to shorten the interval between foraging onset and delivery of rewards for foraging behavior.

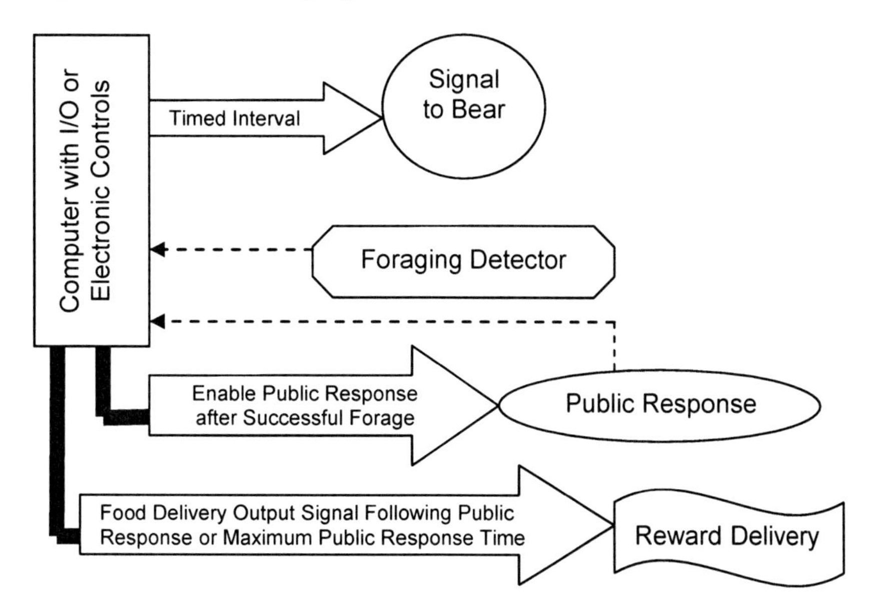


Figure 7-4. Bear is Rewarded for Foraging in Area Where Acoustic Prey Are Heard

Diagram 7-5 shows one logical sequence that can be programmed to provide this kind of enrichment with visitor initiated foraging opportunities. Notice that in this sequence if the public fails to respond the program will eventually initiate the foraging opportunity.

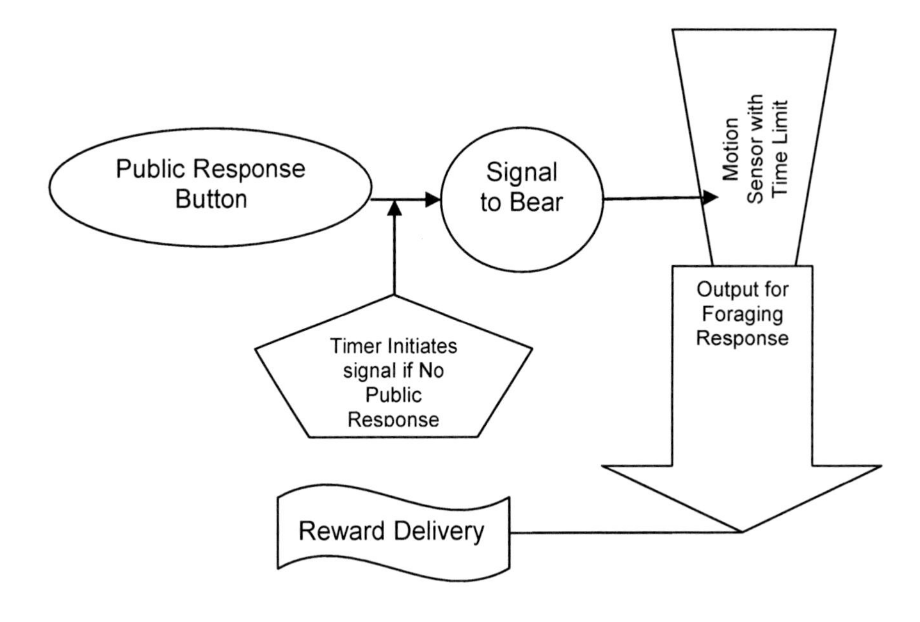


Figure 7-5. Visitors Have Opportunity to Initiate Signaled Foraging Opportunities

Whether you select one of the approaches shown in these diagrams or a program of your own design, it may be wise to use an inexpensive computer as your controller. Simple alterations in a computer program allow you to make outcomes more like those in nature. For example once the bears have learned the relationship between foraging and the possibility of treats being delivered, you may wish to make foraging in response to sounds only lead to success in finding food at random times. This would encourage more active behavior and better simulate the situation for bears in natural habitats where foraging is not always successful.

Bears are among the most often exhibited of zoo animals, and they are capable creatures that deserve active opportunities to engage in the control of some important life activities. Their physical health will also be benefited by rewarding them for efforts that require vigorous activity to find food.

# 8

# MONKEYS

When I was young, a comedian named Lehr made famous the saying "monkeys are the craziest peoples." In retrospect, he was clearly wrong for at least two important reasons. First, his examples almost always involved apes rather than monkeys. And, if he just meant "primates" he was wrong because no other primates have begun to behave in a manner as crazy as humans. One need only look at what we have done to overpopulate and otherwise despoil the planet to confirm this.

At the time, the comedian's saying was one that conjured up for most of us the notion that it was interesting to see what "monkeys" did because so much of their behavior seemed easy for us to understand and a little bit loony. Much of the history of human interaction with captive primates was based on two kinds of experience. As pets they behaved in ways that humans felt they could interpret, and as zoo animals they would respond to the presence of humans.

I remember being saddened by watching an award winning short movie about zoos that appeared about a half century ago. One of the "highlights" that most regaled audiences was watching some children in a European zoo hold up face masks to cause fright responses in young monkeys. Today it is unlikely that such a film would win awards for scenes like this because of the number of individuals and lobby groups advocating greater respect for non-human animals. But a trip to the zoo and observation of human behavior in the primate house is almost certain to bring evidence that not everyone is so enlightened. Banging on windows or cages, making mocking faces to attract or "ape" the residents, and calling names at the captive primates unfortunately remain everyday occurrences.

The literature on enriching the lives of captive primates has grown by leaps and bounds. This is partly fueled by human decency and partly by new federal regulations requiring many kinds of facilities to provide for "the psychological well-being" of non-human primates. There are books, websites and many hundreds of published journal articles that focus on enrichment for primates. It is well beyond the scope of this book to begin to synopsize all of the approaches that have been tried. Instead the concentration in this chapter will be on evaluation of enrichment for monkeys and the presentation of a few recipes. These recipes are based on field experience studying monkeys and efforts that have been measurably effective in improving the lives of captive monkeys.

**How Do You Know That You Have Provided for a Primate's "Psychological Well-being"???**

Many years ago, I served on an advisory committee that met with the folks responsible for enforcing the first regulations requiring that any federally funded research facility provide for the primates' psychological well-being. I found myself repeatedly asking the people assigned this awesome task if they could tell me how assessment would be made to determine that "psychological well-being" was "provided for." In my estimation, this measurement problem still remains. After all that has been done, required by law, and written about enrichment by "authorities," the majority of those involved never clearly identify how they have defined their objective. This is not surprising, unless you believe that we have access to the psyches of non-humans. As discussed early in the present book, I recommend that anyone undertaking enrichment *on behalf of* non-human animals clearly identify how they will *measurably* identify success or failure.

Does "success" in enriching lives of non-human primates mean that they behave more like non-captive members of their species? Does it mean showing fewer behaviors that are unique to the captive circumstance, such as more pacing or somersaulting than is seen in nature for the species? (Erwin & Deni 1979) Or, for primates in captivity, does it mean showing fewer signs of apparent stress than before you enriched their captive circumstance? Unless one defines these goals in advance, the fact that the animals may choose to interact with a particular enrichment activity that you designed is rather weak evidence that you have provided for their psychological well-being.

I do not have great sympathy with the wording of ever-increasing regulations requiring that the "psychological well-being" of animals be provided for, since we do not have access to their psyches. It becomes preposterous when these terms are used in regulations that are now extended to animals where we have no foreseeable means of any access to their thoughts. I would propose the alternative criteria for well-being presented in the next section of this chapter. Frequently I have asked visiting federal inspectors whether these proposed alternatives were satisfactory to meet the legal requirements at facilities in which I work. I have almost always been told something like, "of course that is what I mean by psychological well-being." Armed with this knowledge about what has served to meet requirements for us, I have some confidence that I will not be misleading those of you who might adopt some of these criteria.

Of course you can always do some inquiries prior to spending time and money on an enrichment scheme by asking in advance whether it meets the requirements. You may find that it is not always possible to get a meaningful answer. Many enforcement agency regulations now have instructions for compliance which stipulate that you must meet "locally accepted standards" for enrichment. Once I asked a veterinarian who was a frequent member of inspection teams what he meant by "psychological well-being" so we could do the right thing in establishing our "local standards". He told me that was a silly question and that any fool could tell when a non-human primate was "going nuts" in captivity. Fortunately for my institution, I bit my tongue and just said to myself silently, "I am glad I am not 'any fool'...". I have seen many non-human primates doing things in their natural habitat that we would classify as nutty behavior if we observed the behavior in human primates. There are doubtless things that we do that non-human primates find strange when they observe us.

**Some Suggestions for Possible Goals of Enrichment for Non-Human Primates:**

1. To eliminate negative and potentially self-injurious behavior in the captive circumstance, *even when this behavior is a frequent component of behavior in the wild.*

2. To measurably increase the proportion of species-typical behavior seen in wild conspecifics, and absent or infrequent in the captive circumstance.

3. To give the captive primates something with which they will frequently interact.

4. To provide non-dangerous social opportunities for primates that are social in nature.

While readers can undoubtedly add to this list, I will stop here because I believe the above four items encompass the majority of goals that allow for measurement employed by most folks currently working on enrichment. A careful quick study of these goals will show that while some can be met simultaneously, others are clearly incompatible aims. Additional difficulties lie in selecting criteria for assessment. For example, while pair-housing monkeys may lead to behavioral signs that are more typical of species in the wild such as grooming behavior in macaques, when the groomee has no space to move away from the groomer this may lead to immunosuppression in the animal that is the recipient of this natural behavior.

My own belief is that, no matter how much we love or feel empathy for animals in our care, we do not have any methods that currently allow us to know whether they are mentally well-adjusted. The best we can do is to carefully investigate whether enrichment efforts that we employ meet our intended goals. Primarily this amounts to assessing if measurable effects on the animals' behavior and physiological condition indicate positive or negative outcomes resulting from changes that we have instituted.

I was fortunate in obtaining funding from the National Institutes of Health (NIH) so that we could accomplish a number of empirically oriented studies on various enrichment techniques for use with monkeys. We were surprised by some of the outcomes, and reinforced in our belief that it would not do to simply show that a monkey interacted with something in evaluating whether their behavioral and physiological well-being was improved.

Before presenting some recipes for enrichment, I thought it might be useful to the reader if I summarized some of the outcomes of these research efforts in evaluation of enrichment for non-human primates. More complete descriptions of the research can be found in journal articles and book chapters (e.g., Markowitz & Line 1989; Line et al 1989a; Line et al 1989b; Line et al 1989c; Line et al 1990; Line et al 1991)

1. Simply increasing cage size does not reduce apparently maladaptive measured behavioral and physiological responses to stresses of daily life for individually caged rhesus macaques. This proved true whether the cage was increased to the size required by new federal regulations or doubled in size and made larger than new requirements demanded.

2. Providing even limited means by which the primate could *effectively* interact with the environment measurably resulted in much quicker return to non-stressed behavioral and physiological states following stressful events.

3. Simple non-responsive toys were not used by adult rhesus macaques beyond a very short period. If introduction of manipulable toys is the form of enrichment used, the best method, regardless of age group, is to regularly rotate a variety of such items in and out of the cage, since highest use rates occur shortly after introduction of new items.

4. Social grouping of aged macaques that have been individually caged in the past is potentially lethal.

## Simple Recipes That Have Proven Successful in Occupying the Time of Captive Monkeys

### Simple Recipe 1. Foraging Board

Following the lead of Kathryn Bayne and others (e.g., Bayne et al, 1992) we have used a number of types of foraging devices for attachment to the cages of monkeys. Here is the way to make one that has proven both useful and fairly easy to clean and maintain:

1. Determine how much space you are willing to block in the perimeter of the monkey(s') cage or other enclosure. Build a frame out of a suitable long lasting material such as stainless steel that can be used to keep a piece of dense artificial fleece stretched and tightly held. We have typically made rectangular frames but any appropriate shape allowing you to stretch the fleece will do. (You might envision the frame as similar to the ones historically used to keep wooden tennis rackets from warping.)

2. Drill holes in the edges of the frame spaced in a manner allowing you to easily mount and remove the device from the cage.

3. Purchase or make "j bolts" that are at least ¼ of an inch in diameter and long enough to accommodate the equipment listed in step 4. We have made j bolts out of lengths of stainless steel rod. This involved using a rod bender in the university shop to form one end into a semicircle to hook onto bars or wire mesh elements of caging, and then threading the other end of the rod.

4. After passing the j-bolt through one of the holes in the perimeter of the frame, assemble the following parts on the threaded end of each of the j-bolts in this order: 1) a shoulder washer with a large enough outside diameter to retain a suitable spring; 2) a sturdy compression spring that will allow husbandry staff to compress it in order to install and remove the foraging board from the cage without using tools. The spring must be strong enough so that monkey(s) cannot easily compress it; 3) another identical shoulder washer; 4) A lock nut to retain the "sandwich" (See figure 8-1 below).

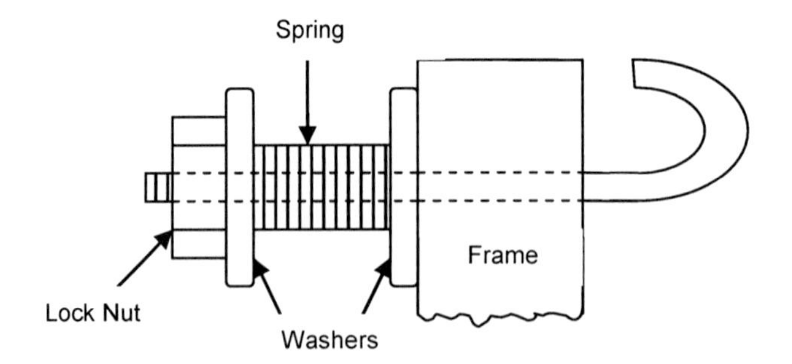


Figure 8-1. Sturdy Quick Disconnect Method for Attaching Puzzles and Feeding Devices to Cages

5. After stretching and clamping the fleece in the frame, load it with a batch of materials attractive to the monkeys and suitable for their consumption. We have found items like sunflower seeds a welcome treat for many primates.

6. If you do a good job of pushing the food treats well into the dense fleece, you will find that monkeys may entertain themselves by foraging for hours.

While the above device is easy to use and fairly inexpensive to manufacture and maintain, it is a labor intensive device for husbandry staff because some of the imbedded materials will inevitably litter the floor when dropped or dislodged by the monkey(s), and because it needs to be removed for loading when the food contents are expended.

**Simple Recipe 2. A Novel Finger Maze**

A number of companies supply various finger puzzles for primates such as finger mazes that can be assembled into different configurations. While we have found them useful in some cases, they also prove to be exceptionally labor intensive for staff since once monkeys become proficient they may empty them in a matter of minutes.

This recipe is for a device (figure 8-2), which was first built by my grad student Cindy Mizuhara and her dad. It has proven to be somewhat more challenging to empty when properly loaded, and occupies primates for a considerable time.

1. Make a "sandwich box" of at least ¼ inch thick Lucite or other strong plastic. This can be done by cutting two rectangles to fit available space on the cage and then making an inside perimeter frame to which to attach these rectangles. The inside dimensions of the sandwich box must be in multiples of the diameter of a ping pong ball.

2. Use a milling machine or precision router to make slots large enough for the monkeys to insert their fingers in the rectangles. For macaques we used slots extending approximately 3 inches in length and with ½ inch spaces between slots to maintain rigidity in the plastic. These slots should be arranged so

that when they are horizontally oriented each row of slots will be centered on the dimension of a ping pong ball, with the balls stacked one above the other.

3. In the other rectangle drill a series of circular holes approximately 1 inch below the top edge. These holes will be used to insert food treats such as nuts into the ping pong ball maze.

4. The frame should be just slightly larger than a ping pong ball in depth. Use Epoxy or plastic welding to join the frame pieces and one of the large rectangles to form the sandwich box. Thread holes in the edge of the other side of the frame so that the other rectangle can be screwed on with stainless steel machine screws to serve as a lid once the ping pong balls are in place.

5. Arrange a method of attachment to the cage such as that described in detail in Simple Recipe 1 above, and illustrated in figure 8-1.

6. Fill the sandwich box with a sufficient number of ping pong balls so that only one space the size of a ping pong ball remains. Screw on the lid.

7. Attach the ping pong ball maze to the cage after shaking the device to randomize the location of the opening in the top row of balls.

8. Load the device with some dry food treats that are unlikely to "gum it up."

These devices have proven to take some time to unload by monkeys, since the manner in which they typically manipulate the balls causes the balls to be rearranged in fairly unpredictable ways. It is also possible to load more than one food treat into the device, provided the food items are not too large in size.

Figure 8-2. Ping Pong Ball Finger Maze for Primate Enrichment

## Simple Responsive Enrichment Devices for Monkeys

The utility of particular kinds of sounds as enrichment for monkeys varies widely between both species and individual animals. Consequently, I strongly recommend that you accomplish some substantial pre-testing of responsiveness of monkeys to the sounds that you choose to employ. Once you find sounds that reliably produce measurable improvements in the behavioral well-being of the monkeys, you may wish to try the very simple recipe for allowing them some control of the sounds presented below.

Some husbandry personnel and researchers have reported that having a television on in the room where primates are housed is calming and apparently entertaining for the animals. If this proves to be the case for the monkeys in your care, the same methodology can be used to allow them control over selected aspects of television.

I have one quick caveat about radio and television programs in animal care facilities. In many cases that I have observed, it is difficult to determine in the absence of systematic measurement whether the primates enjoying the programs are the residents or the husbandry staff.

### Simple Responsive Enrichment Recipe 1. Monkey Controlled Audio

1. Purchase some sound production medium of your choice, such as a radio or compact disk player, that has a rugged remote control.

2. Decide what aspects of the control you wish to place at the monkeys' disposal. For example you may simply wish to allow them to operate the power switch to turn the sound off and on when they wish, or you may also want to give them control over the loudness of the sound (being certain to limit the maximum loudness so that it is not excessive). You might even provide them the means to change selections being played or stations tuned.

3. Place the remote control in an armored box that can be mounted securely to the cage. Drill holes in the box and devise a method to secure the remote control within it, so that a monkey can operate the buttons that you have selected for their use but not destroy the control.

The sound source can be placed out of the reach of the monkey(s) where it can easily be adjusted or new compact disks can be inserted by husbandry staff.

Of course, if you have a substantial budget, you can improve on this recipe in many ways such as having a custom remote control made that is strong enough to prevent damage by monkeys, and too large to throw out of the cage. Then it can be given to them to tote around as they please. You could provide a universal remote control and give the monkeys an opportunity to select from a number of different sound and video sources placed at safe distances outside the cage. A rugged remote control with a touch screen for selection may be a tempting idea. Unfortunately, if placed inside the cage, these screens tend to get disabled by various liquids and semi-solids that the monkeys smear on them.

**Simple Responsive Enrichment Recipe 2. Local Entertainment**

Where multiple cages are located in the same room, and there are solid partitions between cages, the following is an inexpensive method to allow occupants of each cage individual sound options.

Steps 1-3 are identical for this recipe as in Recipe 1 above, except that you will not want to allow the monkeys to change stations.

4. Use the speaker output connection to do *one* of the following:

A. Supply sound to speakers mounted safely in or near each cage by running conduit-protected speaker wires from the sound source to the speakers. If you choose properly matched speakers and sound source output, it will be possible to drive a number of speakers from a single sound source without adding amplifiers for each speaker. Supply each cage with a safely protected set of controls that allow the cage occupant to control sound level and to turn sound off and on. Either place the local speaker in the same rugged enclosure as the controls provide for the occupant, or use conduit to protect the connecting wires between a separate speaker enclosure and the control enclosure.

B. Use a local broadcast system to supply sounds from the sound source to inexpensive receivers at each local set of speakers. While this second alternative may be a little more technologically challenging, it has the significant advantage that if properly accomplished it eliminates conduit. The local receiver, water-proof speakers, and controls for monkey's use can be mounted within a single strong enclosure.

**Simple Responsive Enrichment Recipe 3. Food On Demand**

Attractive food is universally sought by captive animals for reasons discussed in early parts of this book. Monkeys will be seen to greatly enjoy the opportunity to do something that delivers them food, even when they are not hungry and may not consume it. The following very general simple recipe is obviously a beginning "module" that can be embellished as budgets and time allow.

1. Construct or purchase a food delivery system. We have found belt feeders (chapter 20) especially trouble free and versatile for loading a variety of treats.

2. Build or purchase a very simple control device that allows the husbandry staff to adjust the number of responses required of the monkeys to deliver a single food portion. (This can be done for less than ten dollars by using CMOS integrated circuits and a few other components.)

3. Select some rugged response device for the monkey(s)' use. While touch controls have the advantage of extremely long life, they are somewhat more expensive than simple pressure operated switches. In addition, switches that have some excursion provide some tactile feedback for the monkeys to determine that their response has had effect.

4. Attach the manipulandum that you have selected for the monkey(s) in a convenient place for use by the resident(s). Be certain to enclose it in a manner that protects it from destruction by monkeys and prevents monkeys from being shocked by electrical power passing through the switch.

5. Run necessary connections from the manipulandum enclosure to the control apparatus and thence to the feeder through protective conduit.

6. Place the feeding device out of reach of the monkeys and deliver the food through PVC tubing or other rugged conduit to the monkeys' environment (see fig 8-3).

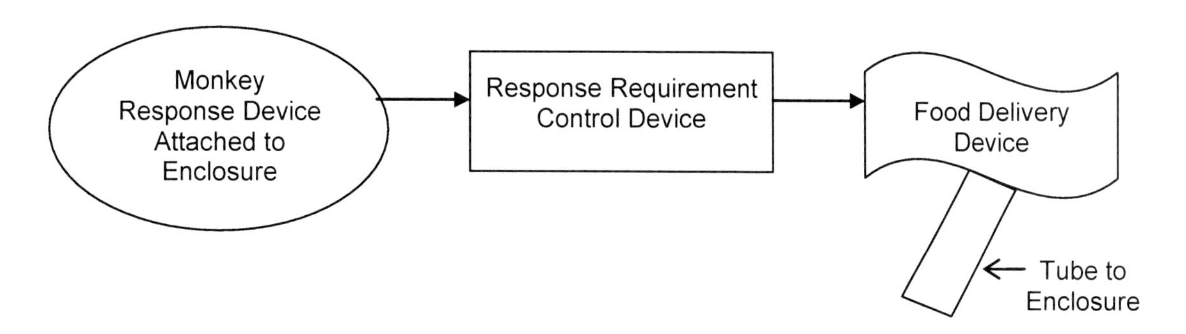


Figure 8-3 Simple Ratio Schedule of Reinforcement Arrangement

An alternative to step five that eliminates the need for conduit is to use a remote controlled relay that is operated by the manipulandum. The remote relay can be placed in the same box as the control unit and used to input response signals to the controller.

## A Few More Complex and/or Expensive General Recipes for Responsive Enrichment

### General Expensive Recipe 1. Custom Built Monkey Controllable Mechanical Devices

1. Build a mechanically operable feeder such as that described in detail in chapter 12 on enrichment for pigs or those in chapter 20. Remember in constructing this device that monkeys, while not as generally strong as swine, have amazing dexterity and apparent passion for disassembling things.

2. Select food such as monkey chow that can easily be delivered by this non-electrical enrichment device.

3. If the enrichment is for environments housing a number of monkeys, install several such mechanical feeding devices in the monkey enclosure so that more than one animal can work to produce food on demand at the same time.

Monkeys are willing to exert substantial effort to turn wheels or accomplish other mechanical tasks that allow them to obtain food. As with other responsive types of enrichment, we have observed that occupation in such tasks often reduces aggression against cage mates and provides more rapid recovery from routine sources of stress.

### General Expensive Recipe 2. For Primates Housed In Cages With Climbing Structures

1. For arboreal monkeys in large enclosures, place motion detectors in various places in trees where climbing is safe for healthful exercise. Protect these motion detectors in securely fastened enclosures that also contain waterproof speakers and the remote relay devices described for use in steps 2 and 3.

2. Connect the outputs of each motion detector to a sending unit for operating a remote control relay in the central response recording and control location for the enrichment equipment.

3. Connect the speaker in each enclosure to a remote control relay that can be operated by the central control unit.

4. Provide one or more feeding devices to deliver food treats to the monkeys.

5. Install a simple computer with an appropriate I/O interface or other control and response recording device of your choice in a protected place convenient for use by husbandry staff.

6. Program the control/recording device so that when monkeys are detected by motion sensors each of the tree climbing responses is registered.

7. Have the program deliver some sound that is likely to attract monkeys, to randomly selected locations where there are speakers and associated motion detectors.

8. Use a system of switches or a simple computer program that allows the husbandry staff to use the computer keyboard to select response requirements for the monkeys to earn food treats.

9. This relatively expensive and complex method of enrichment has the advantage of considerable flexibility in varying the parameters for enrichment. Initially you may, for example, elect to use a program that allows the monkeys to simply climb any tree that currently has the attractive sound emanating from it and deliver food to the closest feeder at the cage perimeter. Or you may choose to have a single such response by any monkey activate a number of peripheral feeders to reduce competition for delivered food and increase the probability that the monkey that has expended the effort to climb the tree will get some of the food treats.

10. Once the monkeys learn that food will be delivered when they climb the trees where sounds occur, you may wish to program the control device so that a random sequence of speakers will produce sounds. If monkeys successively climb to each area where the sound ensues, food treats will ultimately be delivered upon completing the programmed chase. Note that you may add as many locations to this random sequence as you wish, thus offering challenges as monkeys become increasingly proficient at gathering food. You may find that monkeys choose to climb the trees in a timely manner as a group effort rather than depending on one animal to accomplish all of the movement tasks. Finally, note that clever monkeys may choose to remain more arboreal in locations where the motion detectors will always detect their presence. This problem can be alleviated by adjusting the motion detector system or the control program so that responses will only be effective in delivering food when the monkeys *enter* an area rather than when they are statically positioned there.

**General Expensive Recipe 3. Togetherness**

Where you wish to test the potential for safely pairing monkeys or combining them in larger social groups, we have found the following device to be especially useful. It can be custom built to meet requirements for the particular captive facility. For example, in research colonies where monkeys are typically maintained in smaller cages, this device can be built to stretch between cages on opposite sides of the colony room so that monkeys can be allowed access when doors on their home cages are opened. Many zoos, wildlife parks, and rehabilitation facilities have night quarter and holding areas that may also make this arrangement possible.

1. Build a large open-ended mobile enclosure with horizontal bars on the sides to prevent the monkeys from potentially harming husbandry personnel. An example of such a device is shown in figure 8-4.

Figure 8-4. Monkeys Get More Panoramic View of World with Horizontal Bars in Cage Extender

We have found the horizontal bar arrangement to be significantly enriching for monkeys because they can scan the entire room without the strobe effect presented by vertical bars. Monkeys can also more easily use horizontally arranged bars to climb on, thus increasing healthful exercise in captivity. We successfully built the enclosure shown in figure 8-4 out of sturdy 2 ½ inch plastic pipe. However, where cage washers using extremely high temperature water are employed for cleaning the apparatus, such construction may be unwise. Glue used for assembly will tend to eventually give out or the plastic pipe itself, no matter how heat resistant, may eventually warp under these conditions. For such installations I would suggest having the entire enclosure professionally molded out of non-chewable fiberglass suitable to withstand the effects of very high temperature. Where it is affordable, this method is preferable for ease of construction and maintenance.

2. Orient and space the bars that constitute the roof of the enclosure so that dividers can be dropped between them in convenient places to partition the enclosure into various sections.

3. Arrange sturdy vertical guides on the sides of the cage below the locations where dividers may be inserted. Provide a means to secure the dividers when they are inserted.

4. Make a number of durable plastic dividers that can be inserted into the slots that you have arranged. Where possible, make these dividers versatile so that they can also include a means to insert manipulable enrichment devices for the monkeys. The dividers should be constructed so that the monkeys can see through them and eventually touch each other through them. One means to accomplish this is to make dividers with horizontal bars for the monkeys to look between and reach partially through.

5. Insert dividers that are sufficiently separated so that monkeys can see and approach but not touch each other. Place the socialization cage between the cages housing individual monkeys so that one or more monkeys can be introduced into each end of the cage.

6. While the monkeys are physically separated by the partitions, allow ample time for them to habituate to the presence of other monkeys in the enclosure.

7. After you have determined that the monkeys have satisfactorily habituated to each others' presence, insert dividers toward the center of the cage allowing closer proximity than previously. Choose a distance between these new dividers that will allow the monkeys to touch but not grasp or otherwise injure one another.

8. Remove the first two dividers so that the monkeys may now more closely approach each other.

9. If peaceful and uniformly positive interactions are observed at this more proximal distance, remove one of the dividers so now the monkeys can potentially reach each other to groom or otherwise interact. If no conflict arises at this stage, schedule a considerable period of careful surveillance by husbandry staff who can separate the monkeys should they become aggressive. We have usually scheduled observation sessions for a couple of hours at different times of day for at least a week for each of the steps described above.

10. Remove the final divider and allow the monkeys to use the entire enclosure formed by their previous home cages and the connecting socialization device. Be sure to schedule a couple of hours of observation time immediately following removal of the final divider.

In our use of this device we have observed that, when they apparently tire, some monkeys may choose to return to their original home cages. They return to the larger socialization apparatus area when they are more rested. But younger macaques to whom we have offered this opportunity have spent all of their time in the new socialization cage actively looking between the horizontal spaces in the bars and apparently mutually enjoying a view of the entire room that was precluded in their smaller vertically barred individual home cages.

Although no socialization method is without some inherent danger, we have found that this gradual method, coupled with enrichment devices that the animals can manipulate to occupy themselves, is more likely to lead to effective socialization and reduce the potential for fighting. It also allows the husbandry staff to identify animals that are clearly hostile toward one another at stages before they can do physical damage.

In the chapter on lesser apes I suggest some acoustic methods for enrichment that may also be of value for many species of monkey. For reasons discussed in chapter 9, I think that these approaches might profitably lead to combining funded research with enrichment. It would certainly provide more interesting zoo exhibits. In studying the documented behavior of free-ranging howler monkeys you will find much speculation concerning the exact utility of their vocalizations (e.g., Cornick & Markowitz 2002; Gavazzi et al 2008). The captive circumstance might provide opportunities to more easily correlate responses to specific kinds of vocalizations. Seyfarth and Cheney (1980), as well as others, have shown the potential of acoustic playback methods in attempting to decipher what is communicated by specific vocalizations in natural settings. Similar procedures might be used in captivity to identify whether specific kinds of responses to sounds shown by experienced monkeys in the wild are also seen in zoos. It would be especially interesting to investigate the difference in responses to sounds between animals born and raised in captivity and those that have been brought from the wild.

# 9

# LESSER APES

I have always savored any opportunity to observe the graceful magnificence of lesser apes as they brachiate through trees, swinging and leaping to nearby trees while conducting their daily routines. In the wild they forage throughout much of the day, eating fruits and shoots in the trees along the way while they move around and maintain their boundaries. Today there are an increasing number of wild life parks and zoos that have large enough wooded forest domains so that some of these activities can be carried out by gibbons and siamangs (*Symphalangus syndactylus*) in captivity. This greatly heartens me. More than forty years ago, before many of these changes, I began my efforts in behavioral engineering by inventing a method that I thought might be beneficial in enriching the lives of four white-handed gibbons (*Hylobates lar*) that lived in very unattractive circumstances.

One of my students, Brian Johnson, had told me that I would find that Dr. Philip Ogilvie, the Portland Zoo Director, shared many of my feelings about the largely dismal and powerless lives led by most captive animals. Upon meeting Phil, I found that Brian was right. A further surprise came when he and I immediately agreed that our saddest times in the Portland Zoo were spent when we visited the white-handed gibbons. These naturally capable foragers should be able to spend most of their time in trees. Here they were reduced to groveling on the floor for food deposited through a narrow slot in the door of their mausoleum-like cage. Phil invited me to do research in the zoo whenever I wished and challenged me to do something to improve the lot of the gibbons. He apologized because there was no budget for research or for enrichment.

Working partly at home and partly in a lab and shop area that I had built in a university basement, my students and I built equipment that I designed to allow the gibbons to feed themselves high up in their cages. This encouraged them to get some healthful exercise by brachiating, swinging and leaping among aerial bars in their cage to earn food. Even though free food was always provided through the slot in their door at the same time that other animals in the primate house were fed, for more than seven years gibbons in this cage usually chose to work to gather fruit, vegetables, and primate chow at higher areas in their captive quarters (Markowitz, 1982).

Figure 9-1. Harvey Wallbanger Exercises to Get Food

Almost four decades have passed since the appearance of David Chivers' beautifully written and informative chapter about the natural history of siamangs and gibbons in the Malay Peninsula (Chivers, 1972). I would still recommend this chapter as required reading for those planning to accomplish some enrichment for lesser apes. It will provide you some feeling for what they might be doing if not captive, and consequently suggest some fruitful ways to establish whether your efforts on their behalf produce more naturalistic behavior.

The initial recipes in the present chapter are different than any of those previously described in this book in two senses. First, they are steps that I recommend be accomplished sequentially. Second, I have never really had a chance to try this sequence and don't know of any reported results from an effort like this. It is always fun to be the "first kid on the block" to find a new way that is measurably effective in enriching the lives of animals. So, if you have the opportunity to work with captive lesser apes in a sufficiently large exhibit, I challenge you to test this recipe. Be that kid and carefully report the results whether positive or negative.

In addition to knowledge gained from intensively studying published research on lesser apes, my idea for this recipe is based upon some chance observations of vocal behavior of siamangs in the San Francisco Zoo. In spite of the tiny and antiquated nature of the cages for lesser apes and monkeys in the zoo at that time, there were still some natural periodic rhythms evident in vocal behavior for the siamangs. I was marveling one day at the predictable beautiful morning choruses of vocalizations that were exchanged between two groups of siamangs housed at opposite ends of one of the strings of cages. The primate keeper to whom I was jabbering was a good friend and past student. He told me he believed that the reason for these exceptional choruses was that one of the offspring of the siamangs at one end of the string of cages now lived in the distant cage at the other end. With the proper incentive, captive lesser apes can clearly produce awesome calls after generations away from the wild. Many readers will have occasionally heard some long calls of lesser apes in zoos. The chorusing that I have briefly described here was unusual in that it was maintained for exceptionally long periods.

## A Systematic Set of Experimental Recipes for Lesser Ape Behavioral Enrichment

### Recipe 1. Getting Started

It is important to successfully carry out this first, rather labor intensive, effort lest you squander time constructing equipment that may be useless.

1. Carefully choose lesser apes with which to work in evaluating whether this is a potentially useful method of enrichment. Study everything that you can find about the behavior of this particular species. Carefully design a data collection method for systematic behavioral observation. Examples of a few of the categories that we have found useful for systematically studying lesser apes are: brachiation, leaping, food gathering, limb walking, "finger drinking," movements on the ground, and a number of specific kinds of vocalizations characteristic of the species.

2. Make a systematic and lengthy evaluation of the typical daily time budgets for these captive apes. Retain your data sheets and use exactly the same system of behavioral data collection for each step that follows. Be certain to make observations throughout the animals' waking day.

3. Select a commonly reported vocalization that is an important part of the natural repertoire of these primates and is associated with the onset of foraging, social interactions or some other important behavior, but is under-represented or absent in the daily vocal behavior of your focal captive apes. Produce two kinds of recordings in the simplest reasonably high-fidelity manner that is available to you (such as using a good microphone and a quality recorder). The first kind should be of reasonably long vocalizations from the gibbons or siamangs that are the subjects of your observations. The second kind should be vocalizations of conspecifics that are not in this captive group. Depending on what is available to you in the way of budgets and resources, these might be recordings of vocalizations you make of this species in the wild, ones you produce in other captive facilities, or archived recordings which have good quality sound.

4. Now that you have become as informed as possible about the natural history of this monogamous ape species, place speakers at an appropriate distance that might be observed between family groups in the wild. Be certain that you systematically collect behavioral data in the same manner that you did in step 2 while carrying out the following procedure. Try playing a variety of appropriate length vocalizations of both these captive animals and their conspecifics so that you can identify if either or both attract the apes' attention. In doing this use typical natural intervals between sound presentations. Do this a significant number of times, randomly choosing vocalizations for each presentation. Observe whether response to these sounds by the resident apes significantly alters their typical daily behavior.

5. Make further systematic evaluations to see whether moving the playback of the recorded vocalizations enhances activity in the residents or produces no detectable change in the apes' responses.

6. Should there be no significant change in behavior visible as a function of your first efforts in playing species-typical vocalizations, try, try again.....until you find vocalizations that reliably produce desirable measurable behavioral changes.

**Recipe 2. Determining the Best Sound Intervals for Stimulating Behavior**

1. Once you have identified one or more recorded vocalizations that reliably enhance the general behavior of the lesser apes, record the vocalization(s) on a digital chip (chapter 20).

2. Use a computer program or a simple pseudo-random time controller with user adjustable controls that allow you to limit minimum and maximum intervals between playing of the digital recordings. The program or controller should also allow you to adjust the duration of audible sound.

3. Continue using your systematic observation technique to evaluate how frequently such vocalizations may be played on average and still reliably generate desired behavior in the resident apes.

**Recipe 3. Responsive Calls**

This recipe is optional and you will want to test its potential effectiveness with a simple prototype before building permanent equipment.

1. Unless you are truly expert in the use of acoustic devices, you will need to identify an ally who has expertise to help in this recipe. Select voice operated relay equipment and appropriate acoustic filters so that only one type of vocalization produced by the resident gibbons or siamangs is capable of operating a voice operated relay.

2. Connect the output of this voice operated relay by running wires through conduit or by using a wireless remote relay system to transmit information between the voice operated relay system and the playback system that you developed in recipe 2.

3. Use the apes' vocal triggering of the relay to activate the recorded vocal response. Your goal in this step is to see if receiving species-typical responses to their calls will stimulate activity in the residents.

4. Allow sufficient systematic observation time to ensure that your enrichment effort produces more than transient interest for the primates. If there is a diminution of interest, a further step is necessary before progressing from the prototype to more expensive durable and weather resistant permanent equipment. This step would be to evaluate whether mixing in other vocal recorded responses to the apes' calls is more stimulating.

5. If you have a big budget and/or lots of time and energy use multiple voice operated relays and associated filters so that a number of different kinds of calls by the residents will elicit appropriate recorded responses.

It is my feeling that the study of specific reliable responses by apes to particular kinds of vocalizations may be a fundable area of research in zoos. Progressive zoo administrations may find this a desirable way to defray some of the costs of enrichment and to enhance the vision of the zoo as a place to study animals in captivity.

**A General Recipe to Enhance Husbandry and Veterinary Care and Reduce Trauma for Lesser Apes**

There are special problems in delivering both routine and emergency care for lesser apes. Gibbons and siamangs generally do not respond well to direct contact with humans. Lesser apes are exceptionally agile and fast moving and have some nasty weapons available in the form of teeth and fingernails. I have frequently seen unhappy keepers or veterinary assistants chasing lesser apes in attempts to capture them for close examination, vaccination, or other important care. Darting them in emergencies is not easy since they do not have the soft behinds that are the target of choice for darting many species. They are so generally thin and bony that there is always the concern that you may fracture or bruise bones.

Here is an idea that may be useful not only for lesser apes, but also for other animals that present difficulties in capture. This recipe will only reliably serve the intended purpose once the primates learn to routinely use the apparatus. But it has real merit in cases where it is difficult and dangerous to train animals to voluntarily participate in health care by, e.g., presenting their arms for blood drawing or inoculation.

1. Identify some area in the perimeter of the enclosure where an elevated tunnel or cave at least 6 feet long can be constructed with a sufficiently large opening to allow the gibbons or siamangs easy entry. The tunnel does not have to be water resistant and can be made from cage fencing and posts if desired. Arrange cage structures so that the ape can jump directly into the tunnel by swinging from a limb or an aerial bar.

2. Install a feeder belt system (chapter 20) outside the far end of the tunnel so that food can be delivered into this closed end.

3. About a foot inside the tunnel install a door that easily swings in or out but always quickly returns to a closed position when released (in a fashion like one side of an old western saloon bar). The construction and installation of this door will require a skilled craftsperson. The door must be easily operable in either direction with a relatively light push when it is not locked, and it must always reliably return to the same closed position when released.

4. Install a standard magnetic entry detector by placing the permanent magnet on the edge of the swinging door. Then secure the magnetically operated switch in a tamper proof area where it will be operated whenever the door is pushed open inwards. Run the leads from this switch through protective conduit to the area where the controller will be installed in step 10.

5. Locate a good place to install a remote-controlled locking mechanism that will hold the door firmly when it is locked.

6. This step requires that the lock not be activated. Encourage the apes to explore the tunnel by using some easily visible irresistible food treat and placing it near the feeder end of the tunnel. Be patient and go far away after depositing the food so that human presence will not deter exploration of the tunnel and pushing the door open to get to the food treat. *Never lock the door while the apes are in the tunnel during this early stage of the work. Lesser apes are clever and wary animals, and successful use of this new procedure may be impossible if they are locked in this space before entry and exit become routinely established.*

7. Select a method to alert the apes that food will be delivered should they enter the tunnel soon. For example, play the recorded sound of a conspecific, and have the sound emanate from the tunnel.

8. Play the sound or use whatever other signal you have chosen to attract the primates and, if they enter through the door of the tunnel, immediately deliver some preferred food.

9. Continue this procedure at least an hour a day or until your available daily food treats are expended. Do not go on to the next step until you are certain that the apes have clearly learned that attention to the signal and entering the tunnel will gain them access to food.

10. Automate the system by using a computer program and inexpensive computer with an I/O board or a manually adjustable industrial controller. Husbandry staff should have controls to: 1) select a range of pseudo-random time intervals between signaling feeding opportunities; 2) adjust the time that is allowed for the gibbon or siamangs to enter the tunnel to gather food before this particular chance ends; 3) adjust the duration that the motorized food belt moves so that one portion of the food currently in use will be delivered each time. The switch that controls the door locking mechanism must be available in a place where the staff can observe whether the ape is safely inside the tunnel with the door in the closed position. On those infrequent occasions when it is desirable to restrain the animals in the tunnel for purposes of examination or providing health care the switch can be used to turn on the electrical lock. (A more versatile alternative is to employ a hand held battery operated remote control and a remote relay to operate the electrical lock.) The general sequence produced when the system is turned on is as follows: A signal occurs at random times and, if the gibbon enters the tunnel through the door to gain access to the food within a pre-selected time, entry triggers operation of the feeder belt for this foraging effort. If no timely response is made by the ape, another interval ensues before the next signaled opportunity (See figure 9-2).

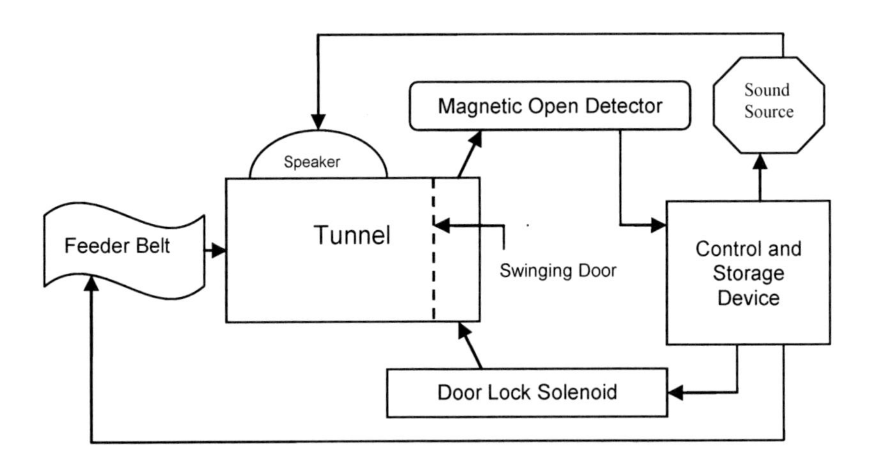


Figure 9-2. Arrangement of Components in System Designed to Facilitate Painless Restraint of Lesser Apes

11. Following those occasions when apes are restrained by activating the lock, they may be wary of reentering the tunnel once the door is unlocked and they have made their way out. Do not be impatient. Allow whatever time is necessary for them to become less wary and again be enticed by the chance to gather food treats.

Equipment and exhibit designers will undoubtedly find more attractive and efficient ways to accomplish the steps of this enrichment procedure than I have described in this recipe. The recipe can be enhanced by making entry into the tunnel just the terminal event in foraging that begins in other areas when signals occur in various regions of the exhibit.

## Some Brief Suggestions for More Simple Enrichment for Lesser Apes

In geographical regions where gibbons or siamangs are exposed to very cold temperatures, we have found that heat lamps are attractive to the animals. The lamps must be installed in a manner that prevents the apes from directly touching excessively hot surfaces such as the lamps themselves. This method will conserve electricity and allow the apes to control when the lamps are on: Focus a motion detector on the area where they can perch to gain warmth from the lamp. Their presence in the area can then be used to turn the heat lamp on and, if you wish to, you can limit the time heat remains on after each entry to the area.

Obviously where enclosed areas are large enough and climate allows, it is ideal to have appropriate real fruit bearing trees and abundant forest vegetation allowing the lesser apes to forage in a totally natural fashion. But even in the case of the largest current exhibits, trees soon get over-browsed by the gibbons or siamangs. In the wild they would forage over more extensive forest areas allowing the trees time to recover.

Many apes are regularly brought into night quarters. During the time they are off exhibit you can spread their daily provision of fruit and browse in elevated locations. This will increase their movement and exploration when they are released into the exhibit.

Where budgets are miniscule and exhibits are antiquated, one band-aid approach that will enhance the lesser apes' lives is to obtain sturdy cargo netting and hawser ropes from a surplus facility or a donor. Sturdily fasten these in a fashion that allows the apes to brachiate and to otherwise exercise by hanging from the bottom of the cargo nets or bouncing on them. It is critically important that you do not fasten these materials in a manner that allows the animals to injure themselves by strangulation or by falling long distances to hard surfaces from improperly secured nets.

Learning about lesser apes is endlessly fascinating. You will not be bored while doing the library and observational research prerequisite to meaningful design of enrichment for these primates. In addition to the variety of color phases and patterning that distinguish some of the species of gibbons, there are some species-specific characteristics of behavior. There are also things that are generally true of all lesser apes and some that are predictable based on genus. For example, all lesser apes tend to be monogamous and many times they will mate for life. In addition to the amazing inflatable throat sac that allows siamangs to produce such high intensity vocalizations, another difference from gibbons is the fact that siamangs tend to show great parental care by males. Male gibbons frequently ignore entreaties by youngsters or ward them off when they too enthusiastically approach. Many of these tendencies are remarkably persistent in captivity. Even in relatively impoverished captive environments, male siamangs carry around young after these offspring are no longer nursing.

# 10

# GREAT APES

There is much written, shown in movies and on television, and debated in various venues concerning great apes. There is a good chance most readers will have encountered important newer ape lore in the time between writing and publication of this book. This, along with the fact that a literature search will reveal hundreds of ideas that have been tried for enriching the lives of apes, makes it beyond the scope of this book to provide a synopsis of past efforts on behalf of chimpanzees, gorillas (*Gorilla sp.*) and *orangutans* (*Pongo sp.*). Enrichment workers new to studying great apes can find many informative and entertaining ways to begin to garner general information, such as looking at Terry Maple's excellent books and chapters about great apes (e.g., Maple 1980, Maple & Hoff 1982). My students and I have also found it very useful to begin with the earliest presentations about great apes that appear in NOVA and other media presentations and watch how the information evolves as time passes.

Your DNA shows that you are the most closely related to those hair-deprived humans. So does your bizarre behavior!

It concerns me that there are so many excessive generalizations about the behavior of particular ape species or genera, and that the information still being disseminated is often quite wrong. I believe that this is partly perpetuated by the fact that the media target audience is in search of simple generalizations. This is apparent in the uninformed nature of many questions about mammals such as: "Which are the smartest animals?," "Who is smarter chimpanzees or gorillas?," and "Are dolphins really smarter than people?" Popular presentations often try to address such questions in entertaining ways, while never telling readers or viewers what criteria are being employed or how their terms are defined.

For all species of great apes, a growing form of improved husbandry techniques involves the systematic utilization of operant conditioning in training of animals to assist in their own health care. Because of the great strength of apes, there have historically been many cases in which tranquilizers were used before carrying out even routine health procedures. There were also many incidents in which husbandry or veterinary staff members were injured while assisting in such common routines as inoculation or drawing blood. Today it is not unusual to see a competent veterinary technician and keeper working together with an ape that has been trained to tolerate the discomfort associated with inserting a hypodermic needle. These animals have learned that if they participate willingly, the discomfort is neither terrible nor long in duration, and that cooperation may lead to kind words and other treats.

It is also possible to arrange behavioral enrichment in a manner providing greater opportunities for examination or easy restraint. For example, some of the games and opportunities to watch videos described below bring these animals into predictable locations. These locations can be chosen so that observation by veterinary staff is facilitated. Zoo veterinarians have often told me how much easier it is to diagnose illness in animals that are actively engaged in enrichment activities compared to those captive animals for which no frequent behavioral activity leads to rewards.

## Chimpanzee Enrichment Ideas

### *Very* Simple Recipe 1. (Good for any Species of Ape)

If the environment is naturalistic appearing but not biologically productive enough to provide sufficient growing food stuff for the animals, it will prove of considerable benefit if the husbandry staff place a variety of food treats in various locations within the exhibit. This will encourage activity and engage the animals' interest for considerable time while they search for food. In many cases this can be facilitated by boring small holes in trees or other areas in which to hide the food morsels.

This has been an apparent source of enjoyment for apes everywhere that I have seen it implemented.

### *Very* Simple Recipe 2. Ape Art

If you have sufficient volunteer staff to work with young apes and well trained keepers who can be on hand to ensure safe control of these animals, you might want to try efforts like those by the chimp-enrichment crew in the Portland Zoo nearly 40 years ago.

Chimp artists were given non-toxic paints and canvas. We were amazed and amused by how much apes apparently enjoyed painting great strokes on canvas, often using their faces rather than their hands to apply the material. As a fund raising device to support this project, a bank manager agreed to display a number of the chimp paintings. Much of this work sold very quickly. One of my friends regaled me by putting a chimp painting in a prominent place and never telling his art lover friends who admired it the identity of the artist......leaving them to praise this anonymous "up and coming" talent.

**Simple Recipe 1. Termite Hunting**

A number of zoos have made artificial "termite mounds" that are loaded with some substance such as honey. Chimpanzees are provided with twigs or other implements that they can use to dip out some of this sticky treat. Often this enrichment effort is accompanied by a nice graphic that tells visitors about the natural ability of chimpanzees to use tools to cleverly gather food such as termites. When I have asked keepers or other zoo staff members why this opportunity was often "temporarily unavailable" for use by the chimps, the answers have almost always centered on two difficulties. The first was that it was necessary to open a door (with mechanisms frequently broken by the chimps) in order to clean out a mess of sticky goo that spilled during dipping honey or other treats. This usually happened because sticky stuff slopped from a hidden open vessel into which the chimps were dipping their utensils to gather treats. The second complaint had to do with the fact that it was hard to keep the whole thing decent looking and still functional. Here is a rather simple way that I invented to make easily maintainable and cleanable "termite mounds" that do not require any doors or other means of entry.

1. Make a suitable lattice on which to place several layers of concrete and a final layer of concrete and vermiculite mixed with the earth on which the termite mound will sit. We have made the lattice by bending rebar into semicircles and then placing a layer of hardware cloth on this strong frame to allow the application of the concrete.

2. Before applying any of the heavy concrete coating, obtain some very durable ¾ inch reinforced hose and form a couple of lengths of it into semicircles that extend from one side of the termite mound to the other, dipping down to ground level in the center of their length (figure 10-1).

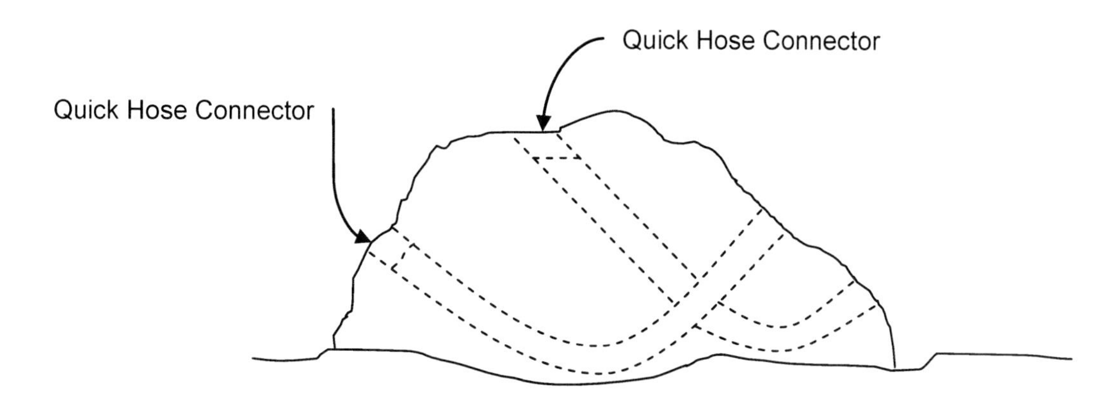


Figure 10-1. "Termite Mound" foraging device

3. On one end of each hose place a simple brass hose terminator (open at the end) so that this will be flush with the surface or just beneath the surface of the final coating of vermiculite and concrete and secured by the underlying concrete layers.

4. On the other end of each hose place a brass quick-connect fitting. These fittings are available at good hardware stores and allow you to easily snap together and take apart hoses while providing water tight connections.

5. Carefully use tape to mask all the hose ends making certain that it still will be easily possible to make attachment to the connector ends of the hoses after applying the concrete. Apply the concrete layers and do a very careful job with the final layer to ensure that the termite mound will blend in well with the setting.

6. Either use very stout bolts into the substrate or adequately bury the base of the termite mound so that it cannot be dislodged by the chimps.

Now you have a "termite mound" that can be made to look quite naturalistic and can be very easily cleaned (figure 10-1). Honey or other desired treats that are suitable for fishing out with long twigs can simply be placed in the two segments of hose by pouring the substance in and letting it rest in the lowest parts of the semicircular hoses. When cleaning time comes, a high pressure hose with a matching quick disconnect fitting can be snapped on to the hose fittings at the surface and water can be used at high pressure to blow out the residual materials from the hoses in the mound.

An added advantage of this design is that more than one animal can forage for treats at the same time.

## Some Thoughts About Sign Language And Ape Enrichment

I have watched seemingly endless hours of video presentations concerning the debate about whether or not the efforts of researchers such as Alan and Trixie Gardner, Roger and Debbie Fouts, and Duane and Sue Savage Rumbaugh have *really* taught chimpanzees to use language to communicate with humans. Herb Terrace has been the principal protagonist for the position that no "true language use" has ever been demonstrated in non humans.

The arguments are sometimes presented in reasonably balanced form and opponents given the opportunity to attempt to rebut each others' criticisms. However, very few of these lengthy popular media presentations address the naivety of searching for "real language" without clearly defining what is meant. It is not surprising that many casual viewers find it laughable that anyone would deny the enormous number of clear demonstrations that great apes can reliably associate American Sign Language (ASL) words with objects and actions. What is missing is sharing with viewers that no one denies that this is an ability shared between apes and humans. The scientific debate has to do with proper syntactical language usage and interspecies exchange of concepts beyond labeling or rote learning of sequences.

A couple of the recipes for enrichment provided below were derived from experience with chimpanzees in an enrichment project in the Portland Zoo that included training young chimpanzees to use ASL. Among the most interesting observations made in this project were that the uses of some of the signs were learned by a second chimp with the "teacher" being the chimp who had been taught the "meaning" of these signs by humans. Even if you are not interested in using methods such as those described below to aid in your efforts to explore questions about non-human language development, providing animals the power to use signs to control some part of the environment is fun. In our own projects I have witnessed apparent joy in apes and dolphins effectively involved in symbolic interactions with responsive humans. This behavior suggests to me that these efforts are enriching for both non-human and human animals.

### More Complex Recipe 1. A Teaching Machine for Apes

1. Mill rectangular or circular holes in a 1/8 inch thick plate of either aluminum or stainless steel using the configuration shown in figure 10-2 below. A good size for the plate is about 12 inches wide and 16

inches high. The three holes in the top row should be exactly the same size and aligned with each other with about an inch of space between each opening. The two holes in the second row should be spaced one inch apart and centered about one inch below the top row. The two holes in the bottom-most row will be used to deliver solid food treats and liquid treats. These openings should be at least 3 inches below the second row and separated by about 4 inches.

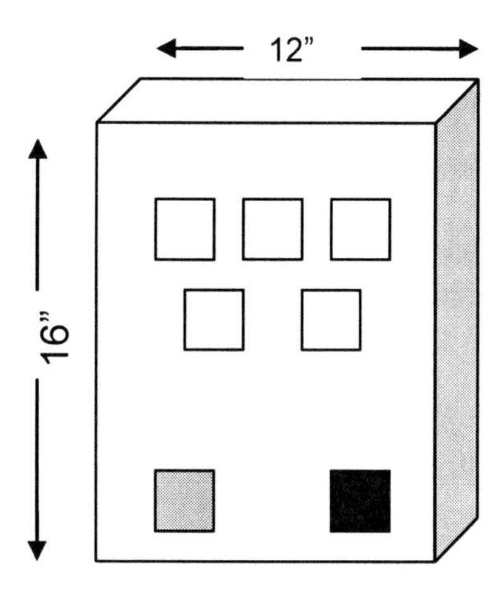


Figure 10-2. "Teaching Machine" for ape sign language

2. Mount or weld four sides to the metal plate in order to form a rectangular box about six inches in depth and open in the back.

3. Prepare a series of durable plastic mounted pictures, each large enough to fill one of the openings that you have made in the top two rows. One of these pictures should be of a human or a chimpanzee displaying the ASL sign for the word "drink." A second should be similar but show the ASL sign for the word "eat." The remainder of the pictures will be of various suitable edibles and drinkables, plus matching pictures of the ASL sign for each of the selected drinks or solid edible items.

4. Arrange a method for quickly attaching and detaching the pictures so that they can be individually seen through any one of the openings in the top two rows. Some sort of clips or Velcro attachment may suffice for your initial efforts.

5. Either find a volunteer with excellent training in making ASL signs or become fluent yourself in ASL so that you can teach the use of signs to the chimps. Begin by teaching apes the sign for some fruit such as an apple and using any of the successful methods that have been used in ASL research with chimpanzees. For example, you can "mold" the fingers of the young chimp into the proper sign the first few times and demonstrate that forming this sign leads to a bite of apple. Progress patiently with verbal encouragement and small bites of apple until the chimp makes the sign without your assistance. It works well to praise loudly and give a whole juicy apple for the first unassisted correct signing. Once the sign for apple is reliably formed by the chimps, teach them the sign for "eat" and reward them with another bite of apple. With some further effort you will eventually be able to teach the chimp to sequentially sign "apple" and "eat" to earn a bite of apple.

Repeat this procedure by teaching the sign for "juice" and giving the chimp a drink of juice for his or her efforts. Then teach that "juice" and "drink" may be combined to earn a swig of the juice.

6. Now you are prepared to try some ASL vocabulary building with the manual device that you have made. Start with what the chimps already know: a picture of the sign for "apple," a picture of an apple, a picture of the sign for "juice," a picture of juice, a picture of the sign for "drink," and a picture of the sign for "eat."

Randomly select whether the picture of the apple or the picture of the juice appears in the center top opening. The left or right opening is used to display a picture of the correct sign being given for the item displayed. Display the incorrect sign in the other opening. Reward the chimp for touching the correct sign by giving her or him a shot of juice or a bite of apple through the appropriate bottom row opening.

7. Once the chimp has mastered this connection and become proficient at getting juice or apple, require the additional proper selection from either "eat" or "drink" that appear on random sides in row two.

8. If you are careful and allow frequent breaks for other activities, you may be surprised to find how quickly young chimps learn how to get favored treats. Be certain that the chimps are proficient in obtaining the first two kinds of rewards before you begin step 9.

9. Now introduce pictures of some signs and associated items that you have *not* manually taught the chimps. Note that what you are asking the chimp to do is learn new vocabulary from the device, rather than by teaching her or him in some traditional manner such as "molding," demonstrating, or shaping.

Although I was surprised to see how quickly some young chimps caught on to learning signs in this fashion, it was not entirely a revelation. In his very first experience with ASL, one of these chimps had been manually taught the word drink by being rewarded with a sip of coke. He was seen the very next day demonstrating to his cage mate the "drink" sign before drinking water from a source at the wall of their enclosure. The female learned her first sign, "drink" from this other young chimp!

Of course, all of these recipes can be used with other species of apes. (If you have the good fortune to work with pygmy chimps you may be astonished at the learning ability of this species.)

***Very* Complex Recipe: Automating the Teaching Machine**

If you are pleased with this first system and it gives some apparent pleasure to the young chimps, the next step is to automate the system. Then they can use it whenever they please. It will take considerable time, expense and effort to do this appropriately if you wish to make the device constantly available to the apes. Only the general concepts are provided in this brief recipe. Following the recipe is a less complex and costly alternative that requires human presence to protect more fragile apparatus.

1. Arrange projector systems capable of displaying the pictures on transluminated touch control screens installed in each of the windows.

2. Use a computer based program to synchronize randomly displayed items and the correct associated buttons to be touched.

3. Install a number of food and drink delivery devices so that each is capable of delivering its contents into the appropriate bottom opening.

You will find this recipe a demanding challenge even for experienced craft persons and programmers. Nevertheless, I have found that challenge is what sometimes intrigues talented folks the most. They may invent amazing ways to accomplish the task.

An alternative method, which is in many ways a more practical and efficient approach to this complex equipment, is to use a large screen television monitor equipped with a touch screen. All of the signs and pictures of items could be displayed in various parts of the screen, and the ape could learn to touch the proper signs to receive appropriate rewards from dispensers. You might find video game designers who would find this a simple but interesting challenge for their talents. The major obstacle would be ensuring that the monitor could withstand occasional abuse by the apes. In most cases this will require constant human monitoring when apes have access to the touch screen display.

I am currently working on the design of "apps" for electronic tablets that will allow non-human primates the chance to entertain themselves by choosing any of a variety of audio/visual options. In first application trials we will give them the opportunity to choose between soothing sounds of friendly people who care for them, movies of conspecifics in natural surroundings, etc. Of course we will have to evaluate whether these materials are effective in giving them apparent enjoyment, or whether any of them may be disturbing rather than enriching for the primates.

One could doubtless use well made devices that are rugged enough to withstand any abuse by primates to do things like speed up training of ASL and to employ in research on audio and video discrimination abilities. I feel sure someone must have thought about that long before me.

## Orangutan Enrichment

I personally greatly favor finding ways for captive orangutans to inhabit forested areas where they can spend considerable time in sturdy mature trees. But there are many attendant problems to this approach for husbandry staff. Providing arboreal opportunities for orangutans is not an easy matter. (Markowitz & Spinelli 1986; Markowitz & Eckert 2005; Markowitz & Timmel 2005). Chapter 20 includes some discussion of problems that arise when planning enrichment for this endangered species.

Figure 10-3. Safe Transportation for a First Forest Visit

Here are a couple of simple ideas for enriching the lives of orangutans in more restrictive captive environments:

### Recipe 1. Cinema

This is a recipe that we first successfully employed more than 35 years ago. At the time there was a general belief that it was a terrible idea to put an adult male orangutan in a moderately sized enclosure with a female and their very young offspring. On the basis of very limited field studies, it was thought that male orangutans were essentially "rapists" who only came into contact with females when they wanted to mate, but otherwise rejected contact with conspecifics. Having watched the gentle and often affectionate interactions between a long-mated pair of orangutans, the zoo director and I decided to carefully try allowing them to remain together along with their infant offspring. We were vigilant and prepared to separate them with tranquilizer darts if necessary. To entertain them during this effort, I designed and we built the system described in the following recipe. Figure 10-4 is a photograph of this family group cuddled with their arms around each other while watching a movie.

1. Install a large translucent screen so that it easily can be viewed on one side by the orangutans and on the other side by visitors and husbandry staff. Be certain that the screen cannot be reached directly and damaged by the orangutans who delight in taking things apart!

2. Install a motion picture projector or other video projection device in a safe place outside the exhibit where it can be focused on the translucent screen and protected from damage by zoo visitors.

3. Select a variety of films that you think might be entertaining for the orangutans.

4. Observe the orangutans' responses to the video presentations and adjust your program selections to include those that most hold their interest and entertain them.

Figure 10-4. Orangutan Family Night at the Cinema

**Recipe 2. Movies on Demand**

1. Repeat the steps in recipe 1.

2. Armor a push button switch so that the orangutans cannot disassemble or otherwise destroy it. Attach this switch to a solidly protected remote control that allows the orangutans to turn the video display on and off when they wish to do so.

**Recipe 3. A Remote Multifunction Control**

If you have a bigger budget:

Repeat Recipe 2, but this time install a number of armored push buttons in a sturdy console that allows the orangutans to choose between program content and sound levels for the video presentation as well as to turn it off and on. This will require the availability of a number of video sources or use of a system with lots of digital memory available that allows simple switching between various recorded materials.

**Recipe 4. An Idiot Box for Our Relatives**

Place a television set with a large screen in a protected place where the orangutans can view it. Provide them appropriate armored buttons to use in turning the set off and on, changing channels, and adjusting sound volume.

**Recipe 5. Puzzles to Solve**

Orangutans are the "engineers" of the great apes and greatly enjoy taking apart most anything on which they can get their powerful hands.

This simple and inexpensive recipe is rather labor intensive and requires some significant rapport between staff and the orangs. It also requires an enclosure that precludes tossing of objects at humans. If the orangutans are housed in groups, careful surveillance is necessary to make certain that the disassembled parts are not used as weapons.

1. Build a number of hardwood wooden puzzles out of very substantial parts. The puzzles should be very challenging to disassemble, and as complex as you can afford to build. An optional feature is to place some food treat for the orangutans' pleasure inside the part that must come apart last. Be sure to make component parts that are too large to be swallowed by the apes.

2. Once a puzzle has been taken apart by the apes, retrieve the solved puzzle parts by raking them out of the environment or by some other practical means.

3. Vary the order in which different puzzles are provided for entertainment.

Based on earlier work summarized in my 1982 book, I know that in very restricted environments orangutans are entertained by automated electronic games such as tic-tac-toe. I hope, however, that we are progressing beyond the point where many of these wonderful and rapidly vanishing animals are housed in such impoverished circumstances.

## Gorilla Enrichment Ideas

The following 3 recipes are derived from some rather odd experiences in zoos.

### Recipe 1. Manual Picture Viewing

Shortly after a new large gorilla grotto was completed in the San Francisco Zoo, I toured with one of my students who was also a zoo keeper. She told me that during an early attempt to introduce the gorillas to their new home, the dominant male became interested in some supposedly gorilla-proof plumbing that was used to deliver water into an overflowing pool. He did not know that this interesting looking fixture was "gorilla-proof" and entertained himself by dismantling it. Next we went to take a look at the indoor quarters into which various groups of gorillas were rotated while others were on exhibit in the grotto. On the way in, we both grieved that this expensive new exhibit was designed so that the majority of visitors would look down at the gorillas. We agreed that this was really an unfortunate flaw in the architectural design.

While we were indoors the dominant male came up to the edge of the holding area to visit with this familiar keeper who often fed him treats. I happened to have a new very large calendar with me that had beautiful full page pictures of various animals. On a lark, I stood just outside the bars and displayed the pictures slowly one at a time to the gorilla. Surprisingly, he not only paid careful attention but showed responses to the pictures that depended on their content. He studied with apparent interest many of the pictures of other primates and birds. When big cats or snakes appeared he shrugged his head violently and/or turned away. I thought that it might be more interesting for him if he could turn the pages himself.

1. Pretest a number of large photographs for effectiveness by displaying them to some of the gorillas that you are tending.

2. Once you have selected pictures that you know are of interest to the gorillas, have the pictures laminated in heavy plastic.

3. Affix the plastic laminated pictures to appropriate sized sturdy mounting boards and equally space a number of the boards on the perimeter of a cylinder mounted on an axle. Make a few such devices if budget allows and by all means make many picture boards so that you can provide variety for the gorilla by attaching different assortments to the cylinder periodically.

4. Mount a very stout rotary handle extending from the top of a heavy anodized metal base so that it can be rotated by the gorilla. If you made multiple displays, install one handle in line with each. On the outside of the cage far enough away or otherwise protected to prevent the gorillas from grabbing, mount the cylinder so that the axle descends into the anodized base.

5. Inside the base, install toothed gears on the display axle and the handle shaft. Connect the gears either by means of a connecting gear or a drive chain so that as the gorilla turns the rotary handle inside the cage, the pictures rotate into view. (See figure 10-5).

Figure 10-5. Gorilla Turns Handle to Select Pictures

**Recipe 2. Picture Automat**

If you find that this device proves to be of continuing interest to the gorillas, you may wish to enlarge the scope of things that they can display to themselves. This recipe describes only one method to enhance the scope of control for the gorillas and I am sure that readers will invent interesting alternatives.

1. Place two pairs of sturdy touch controls where they can be operated by the gorillas.

2. Run the connecting wires to a slide projector in an armored case outside the gorilla cage in a manner that protects the projector from destruction.

3a. If there is an appropriate area of the enclosure area that is dimly lit, install a sturdily framed translucent screen in a manner that will allow the gorillas to view the image from within the enclosure and humans on the outside to view the image as well.

3b. If all of the enclosure is well-lighted, make a shadow box arrangement outside the cage so that projected images will appear on the enclosed translucent screen and place it in an area easily viewed by the gorillas when they are resting at the touch control location.

4. Rewire the video device control circuits to allow the gorillas' touch controls to activate the on, off, advance, and reverse functions.

Now you have built a system that will allow you to easily provide a variety of pictures for the gorillas to display to themselves. Where devices like this have been used in the past, I have observed that husbandry staff gets much enjoyment out of bringing their own favorite video materials to see what effect they may have in attracting the interest of the apes.

A slightly more complex contemporary alternative that has many advantages will require more extensive protection from visitors if used in a public area. An inexpensive computer and large screen monitor will provide much versatility and require less service than using a slide projector. Using this equipment, one can easily store a vast number of pictures in the computer for playback on demand by the ape. Another obvious advantage to this approach is that it allows a great variety of alternatives for enrichment such as allowing the apes to select motion pictures. As I do final editing of this book, the price of plasma and LCD screens continues to plummet, making this a more affordable method for institutions with limited budgets.

**Recipe 3. Allowing the Gorilla to Show Exceptional Strength**

Some time ago I was watching with interest some interplay between a bellicose zoo visitor and a gorilla who was housed in a very small cage. Their interaction was by means of each holding on to the end of a very heavy rope which passed through a firmly anchored pipe that was part of a railing system. At first it appeared that the young man was winning a tug of war, and that the gorilla was nonchalant about losing the contest. Suddenly the gorilla, who had obviously been observing the visitor's behavior closely, used his superior strength to give one mighty tug. He pulled the amazed braggart into the railing, causing him a minor injury. Of course the contest opportunity was soon removed for fear of law suits from injured guests. Memories of this event lingered with me for a number of reasons. While the chance for a tug of war with visitors was available, the gorilla showed some continued interest in this chance for interaction. This was surprising since gorillas so often prefer to turn away rather than interact with unfamiliar humans. Humans were quite impressed when they found that even when two or three burly men pulled simultaneously on their end of the rope, the gorilla almost always won contests. As you might expect, many more people gathered around this exhibit than other areas of the zoo and spent considerable time watching and learning about the gorilla's strength.

The following recipe is only offered as a means to give gorillas in impoverished exhibits something to do and to attract visitors' attention to the strength of these wonderful animals. As budgets allow, it is a much better plan to arrange for gorillas to have ample space that includes places to gain privacy, opportunities for social interaction with conspecifics, and access to natural foraging materials. There are a dwindling number of gorillas in the wild, with today's best estimate for mountain gorillas being only 650 individuals living outside of captivity. One would hope that humans wishing to conserve this species in captivity would be able to find funding and construction talent to provide much more adequate living quarters than the iron-barred concrete arenas in which many gorillas still reside.

These steps are intentionally very general because the exact method of installation will depend on the construction of the exhibit and the visitor viewing area. In some cases, in order to efficiently install the equipment so that it will be safe for use, it may be necessary to temporarily remove and then reconstruct parts of concrete flooring or walls. .

1. Obtain a pair of pneumatic dampers sufficiently well constructed to cushion the strongest pull that a number of humans or the strongest of gorillas can exert on a lever.

2. Make 2 T-shaped levers out of steel rods that are at least one inch in diameter and make certain to file away any sharp surfaces on which the contestants might injure themselves.

3. Depending on the construction of the cage and surroundings, either:

A. Install the two levers and their associated dampers on opposite sides of a wall and pass a stout steel cable through a hole tunneled in the wall (see fig 10-6); or

B. Install the levers on axles in the substrate adding whatever is necessary so that the contestants can pull on them while standing fairly erect.

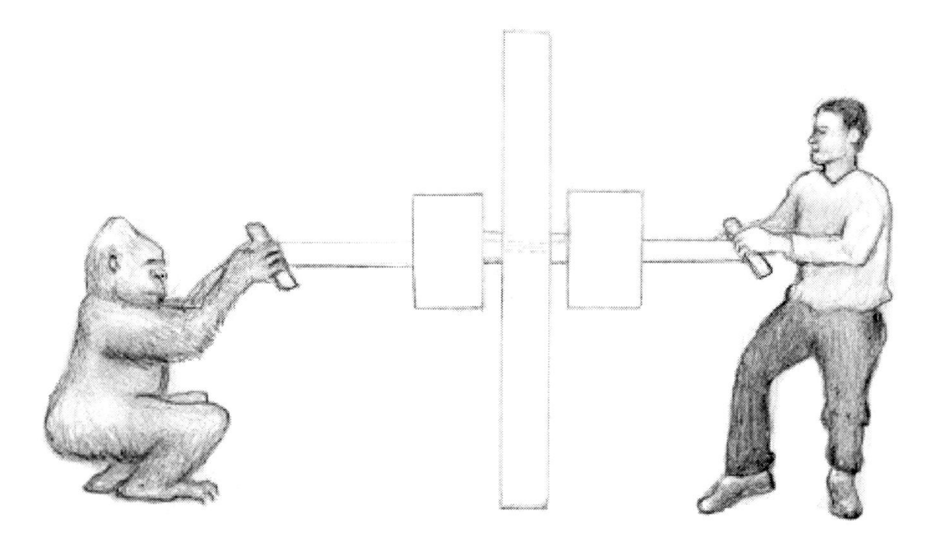

Figure 10-6. Hairy and Hairless Apes Contest in Tug of War

The following steps are optional and depend on greater budgets, so before undertaking them you will want to decide how long gorillas are likely to be housed in this captive environment.

4. Install a strain gage (or a load cell) so that tension on the cable can be translated into a measure of strength that the public will understand.

5. Install the readout for this device in the public viewing area and run the connecting cables in a manner that is protected from damage by the gorillas or visitors.

6. Alter the steel cable by attaching a device that can present significant resistance to the efforts of the gorilla when there are no humans offering competition in the tug of war.

7. Arrange some method of delivering food treats to the gorilla when she or he successfully moves the lever a significant distance to their side. This can be done by placing a magnet on some protected part of the device that moves when the ape pulls the lever and installing a magnetically operated switch in a position where it will be operated when the magnet is moved a required distance. The switch can then be used to operate an appropriate treat dispenser that drops food near the gorillas' lever (see fig 10-7).

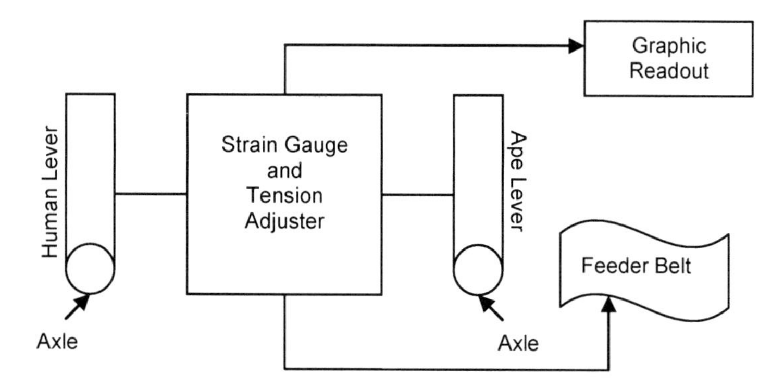


Figure 10-7. Tug of War with Food Reward and Graphic Readout

If you are fortunate and have the opportunity to spend quality time with any of the great apes, you cannot help but admire and come to love them, unless you are the most insensitive of humans. They are individuals with character and are worthy of our every effort to ensure their longevity in the wild and conserve their habitats. It is my belief that we should provide apes in captivity as much opportunity as possible to exercise their abilities and control some aspects of their own lives.

# 11

# SNAKES

By now you thoughtful readers doubtless have concluded that this book might more properly be titled "Enriching Mammalian Lives." Indeed, I confess that my own limited knowledge about captive creatures other than mammals has led to this excessive concentration. I decided that I would stray from this primary concentration on mammals in the current brief chapter to describe some experiences that may prove useful in your advance planning for enrichment. This chapter also provides opportunity for me to return to one of my favorite themes: "Reaffirming That Animals Are Smarter than Investigators" (Markowitz 1982 pp 132-147). The first two cautionary recounted experiences will help confirm that *Homo sapiens* are the silliest and most dangerous of all mammals.

**An Experience with Humans and Snakes in a Natural Setting**

Everyone truly interested in animals has learned that stereotypical ideas about snakes have led to serious mistreatment and misunderstanding. Most every zoo that has hands-on demonstrations by staff uses some relatively harmless species of snake so that humans can ascertain that snakes are neither slimy, nor necessarily unfriendly when properly handled. Today it is not uncommon to encounter someone who has chosen a snake as their favorite pet. Unfortunately, the capture of snakes for the pet industry has further increased our already excessive collection and decimation of some species for food and other human uses. One of the most important areas of current concentration in conservation education should be teaching "eco-tourists" how to cause minimum disturbance to wildlife that they are privileged to observe.

Our seven years of research in the forests and jungles of Belize were begun with the substantial support of Oceanic Society Expeditions (OSE). Under the leadership of Birgit Winning, this organization solicited paying volunteers to help support the research effort. In return we trained the volunteers to help in recording data on black howler monkeys (*Alouatta pigra*). As part of the enrichment for visitors who supported our research, OSE provided naturalists who taught the visitors about a wide variety of local flora and fauna. In addition to inviting questions and teaching as much as we could about local ecology, we cautioned participants not to excessively disturb local wildlife and to protect themselves from some of the potential harms of working in the jungle.

The two relevant bits of information included in these orientations were: 1) how to temporarily handle creatures such as non-venomous snakes, and 2) how to identify dangerous critters in order to do one's best to avoid direct contact with them.

The most venomous of creatures in the forests of Belize is the fer-de-lance (this is a Creole-French name meaning the "evil spirit of sugar plantations"). This lance-headed snake does not actively prey upon humans, but its natural camouflage makes it easy for a forest visitor to accidentally disturb and arouse the snake's defensive behavior. An untreated fer-de-lance (*Bothrops atrox*) bite can be fatal, and if the snake by chance strikes a blood vessel death may occur very quickly unless anti-venom is administered. Included in the lectures by those of us leading tours was the passing around of a large jar with a preserved specimen of fer–de-lance so that there would be no problem in identifying this species of snake should one be encountered.

On one occasion the naturalist had special expertise with snakes. He discussed their natural history at length prior to tours and demonstrated with great emphasis how to carefully handle snakes in a manner where injury was unlikely to occur to snake or handler. He also heartily invited the volunteers to call attention to any snakes that they might see so that he could help in their identification and others would have the chance to see them.

On one tour, an enthusiastic visitor came directly to him carrying a snake which she had just found on the trail through the bush. She handled it in the manner John had instructed and stuck it in his face for identification. Unfortunately, it was a fer-de-lance. She had obviously not paid much attention to that part of the lecture that used jarred specimens to clearly demonstrate species appearances so that snakes could be identified if encountered.

In retrospect, this true story has a happy ending. For those of us who knew the naturalist who led this tour, it even provided some real laughs imagining his reactions to what happened. The naturalist's apoplexy was recounted to us in detail at a later staff meeting, and we were regaled by the story once we learned that no creature was injured as a function of this serious human error.

This account is detailed here as one of the reasons that there is concern that ecotourism may be a mixed bag when it comes to enhancing the lives of animals living in nature. On the positive side, there has undoubtedly been significant improvement of conservation ethics on the part of participants in sensitively run tours. There is also an increased probability that those who have experienced the wonders of exotic natural surroundings may become significant contributors to the establishment of a greater number of natural protectorates. However there is the paradoxical problem that the ever-growing interest in ecotourism has led to increased destruction of some natural habitats as a function of excessive human traffic and intentional or unintentional direct disturbance of animals living in nature. At a minimum, our efforts must include education about this paradox and the need to limit human intrusion into nature reserves.

**A Failed Attempt to Educate and Bring Further Honest Information about Husbandry to Zoo Visitors**

During a dozen summers in which I taught and worked on enrichment in the Honolulu Zoo, I was fortunate in beginning many lifelong friendships with zoo staff. The principal keeper responsible for reptiles was among these friends, and he grieved about the fact that visitors never had opportunity to see the capture behavior that snakes exhibited when they were fed live mice. This daily activity was always conducted out of public view. I suggested that proper graphics and keeper talks to the public about the importance of active prey in establishing vigor in many species of snake would win the day over squeamishness. It seemed to me that we should provide visitors the educational opportunity to witness this behavior.

I lobbied the zoo director to allow us to develop such an exhibit. We provided a plan, including the usual caveats, such as being certain that only small mice were used as prey in order that the role of predator and prey not be reversed. (There are occasional warnings within the zoo community about cases in which larger rodents have been provided as "treats" and have managed to attack the snake rather than vice versa.)

Prey Roles Reversed

The Honolulu Zoo never had a large budget, and considerable voluntary overtime was put in by the reptile keeper and his friends in establishing this viewing opportunity.

It worked! But, I had not counted on the strange sensibilities of some zoo visitors. The family of the head of one of the zoo's largest corporate contributors came to visit the zoo. His wife accosted the zoo director telling him that seeing a poor mouse attacked by a snake was the worst experience of her child's life. She further asserted that she would personally see that her husband's corporation, as well as those of other friends, ceased contributions to the zoo if such barbaric things were available for children to see.

There was also a direct complaint to the mayor, and the new viewing opportunity ended abruptly. I take full responsibility for this fiasco and acknowledge that I expected too much understanding on the part of every visitor who brought their kids to the zoo. For some, seeing animals eat other animals is too horrific even if it is a common behavior in nature and necessary for maintaining the vigor of carnivorous species.

So, where is the recipe?? There are other zoos and theme parks that have mini-lectures by staff and public displays in which live prey are consumed by reptiles and no visitor complaints arise. Before lobbying friends to undertake such an effort in a community where there is significant potential for opposition, I would recommend a public survey. Personal solicitation of friends in the media to make the plan widely known before spending precious time and money would be wise. While this approach is certainly not fail-safe, in retrospect it would have provided my friend the zoo director a better defense when he was called on the carpet by the mayor. I would also advise that reptile exhibit designers provide substantially naturalistic environments in which visitors have the opportunity to witness prey capture and consumption. The inclusion of shrubbery and other camouflage may reduce the tendency of some visitors to think that Mickey or Minnie is being attacked in their bedroom by some sneaky predator.

A talented exhibit designer armed with solid knowledge about the species involved, might design a captive environment in which small rodents or other prey were released into the snake enclosure but with the possibility of escape into some crevice. The prey could be channeled from this escape route back into the release holding area for later reintroduction. This should generate interesting activity for the snake and would certainly appear more naturalistic.

**Snake 1 Researchers 0**

One of my graduate students, Jim Campbell, had great affection for snakes. I remember well the time when his roommate's father was invited to visit them. The visitor opened the closet to hang up his coat and was confronted with a group of rattlesnakes! Seeing this reaction, and hearing about his friend's father's dislike for snakes in general, stimulated Jim to remind me how much he wanted us to undertake a project to teach the public more about the typical behavior of these wonderful reptiles.

Jim taught me a lot about snakes, and I also went to the university library and spent many enjoyable hours learning more. With this preparation I helped him to devise a plan which was supposed to provide zoo visitors the opportunity to observe the thermoregulatory behavior of some of these ectotherms. I have always advocated the publication of negative results as well as those that confirmed working hypotheses. Here is one recipe that will *not* work well:

1. Cleverly design a nice home for a snake, incorporating a heat lamp and programming controls to simulate the changes in temperature that arise as the earth rotates and the sun is seen to move through its daily cycle.

2. Automate a program that allows this apparent cycle to occur several times a day and accompany this with signage for visitors that tells them when "the sun will come up, when night will set in, etc."

3. Make certain that this exhibit has an easy place for the snake to bask in the sun and take advantage of the warmth from rocks that have been heated by the artificial sunlight. Also provide shade so that the snake can cool itself when necessary.

4. Remember that in nature when a snake tucks itself into crevices and under rocks to become less vulnerable to predators, its metabolism drops enough to make it less mobile. This means that you want to be certain to include in your plans a place for such activities.

5. Since you have a limited budget and you want the public to be able to see this naturalistic behavior, you can accomplish the essential part by providing a tube down which the snake can descend when simulated night approaches and from which it can emerge when daylight begins to bring warmth.

If you are as successful as Jim and I were, you will find that the snake *never* totally emerges from the tube but instead occasionally moves its body *very slowly* either further toward the outside or deeper into the tube. In this way the snake can outwit the designers by expending very little effort to maintain an appropriate temperature. There is no need to move into any external place for shade or to even emerge fully to bask in the sun since the snake can thermoregulate by minimal movement up and down in the cylinder.

In addition to providing this as a negative example, I have included it because there are clues about how to more successfully accomplish the original goals. First, if you have a really big budget, you may want to incorporate some nocturnal predators in the same large exhibit area and really abundant rocks and crevices for the snakes to use as protection when artificial night falls. Second, if you fully automate a slow cycle of movement of the artificial sun across the sky, you are much more likely to produce some naturalistic thermoregulatory behavior than we did when we varied the intensity of a fixed position heat lamp.

**Hope Lies Elsewhere!**

Fortunately for the reader, there are many successful ideas for enriching the lives of snakes and providing more adequate stimulation for naturalistic behavior that can be witnessed by visiting a modern zoo home for these animals. Under the leadership of my dear departed friend Dr. Paul Chaffee and his excellent curators, staff members at the Fresno Zoo were among the first to develop a greatly improved reptile house in which much naturalistic behavior could be witnessed. All of the exhibits in this facility had individual solid state controllers that allowed the staff to select thermal cycles for each enclosure. This is one of many innovations on behalf of animals by Paul Chaffee, after whom the Fresno Zoo has been renamed.

For those working with small enrichment budgets for snakes, you can find some effective and very inexpensive approaches in places such as the *Animal Keepers' Forum*. Among the means that have been reported to be effective in stimulating activity in snakes are the use of various odors. Laine Burr reported some work in the Jacksonville Zoological Gardens (Burr, 1997) in which the effect of introducing carnivore smells into the captive homes of snakes was "immense" in eliciting species-typical behavior. Some of the observed responses were alert postures, tracking of the scent, movement away from the odor, tongue flicks, and tail vibrating.

The routine changing of substrates by use of wood chips, dirt, sand, torn newspaper, dried leaves, and other inexpensive or free materials has also been reported to be stimulating for snakes.

A number of keepers have reported the careful use of existing outdoor areas where snakes may occasionally be temporarily released to allow them to experience a variety of substrates and even to swim in shallow pools.

My final suggestion is the incorporation of snakes into large multi-species exhibits. Of course this would require a visitor clientele sophisticated enough to tolerate seeing occasionally successful natural predation in action.

# 12

# PIGS

A good friend of mine was subpoenaed to testify as an expert witness in the trial of some people who had stolen some captive dolphins to release them to the wild. When asked how bright dolphins are, his reply was that they are about as smart as pigs. Some members of the audience gasped at this response. But it was really meant as a compliment by a person who has significant experience in studying, writing about, and providing veterinary care for dolphins. He also knows quite a bit about the abilities of pigs. Pigs are quick to grasp new ideas, to solve puzzles presented to them, and to learn things that bring rewards from humans.

But there are some special problems when it comes to convincing humans about pigs' natural abilities and sensitivities. One problem is that many are unwilling to recognize that the source of their favorite breakfast meat might possess intelligence.

Inevitably the question of animal consumption by humans will divide many of us who are truly dedicated to improving the lot of all living creatures. The following brief account suggests that sincerely caring about the well-being of an animal is not necessarily inconsistent with continuing to be an omnivore.

I was sitting on the front steps of a building near the nation's capital with my friend Roger, awaiting a taxi to take us to the airport. We had just finished providing testimony to a governmental committee concerned with the legal enforcement of new animal welfare legislation. Our private discussion of the week's events was interrupted by a group of very zealous animal rights activists. They knew about the meeting and wanted some assurance that we had included the fact that humans should not be consuming other animals.

After we listened for a few minutes, my friend, whom these protestors recognized as a person who spent most of his life working to enrich the lives of chimpanzees, spoke slowly and carefully. He told them that we both understood their position, but that this did not necessarily require that we agree with their demands. Roger then shared recollections of his experience as a young person growing up on a traditional farm and nurtured by his parents who were animal lovers. The major point that he hoped to share with the protestors was that throughout these formative years he never ate an animal that he didn't respect or care about. One may also see similar ethical views in the traditions of many Native American traditions, which include showing respect and reverence for the animals that provide them sustenance.

Other difficulties in recognizing the intellectual capabilities of pigs may be linked to common pejorative terms such as "filthy pig," and to the frequent appearance of "talking pigs" as silly characters in comic strips and animated movies.

Some young children who grow up watching too much TV may be surprised to discover that pigs do not really stutter like Porky, are not really arrogant and self adulatory like Miss Piggy, and do not spend time heroically helping humans in the fashion Babe does. After spending time with real pigs, only insensitive humans have the misconception that they are ignorant and filthy animals. The more that one learns about their ability to learn new things quickly, the greater one grows to appreciate the intelligence of pigs. This knowledge suggests that these clever animals might benefit from much richer captive environments than those in which we typically find them. There are a variety of kinds of swine and one will have to tailor the following recipes to make them appropriate for the size and strength of the target species. But, it is my carefully considered opinion that most pigs might have some fun with these kinds of additions to their lives.

**Recipe 1. Provide Opportunities for the Pigs to Forage in a Variety of Ways to Obtain Food Treats Whenever They Wish**

1. Identify some foods that are most attractive to the pigs in your care. You will be aided in this effort by the fact that pigs are most willing to consume a wide variety of nutritious and semi-nutritious substances.

2. Remembering that pigs are quite strong, sometimes astonishingly heavy, and have a propensity to dismantle things when they can, add some sturdy elements to their environment to provide some intellectual challenge in obtaining food. The specific elements will necessarily depend on the nature of the pigs' captive habitat and available budgets, but here are a few ideas:

A. If you have sufficient room outside the enclosure at ground level, install a sturdy wheel at the base of the perimeter so the pigs can root in order to winch in food that they can view at a distance from their enclosure.

Figure 12-1. Rooting Device for Use at Perimeter of Enclosure

B. If only primarily vertical space is available outside the enclosure, install a similar rooting wheel, but this time have it operate a vertically oriented clothesline type device which they can bring down to drop food through a shoot into their enclosure. One easy way to do this is to mount the rooting wheel near ground level at right angles to the perimeter fence or wall and to extend an attached axle to the outside to drive the clothesline when the pigs root.

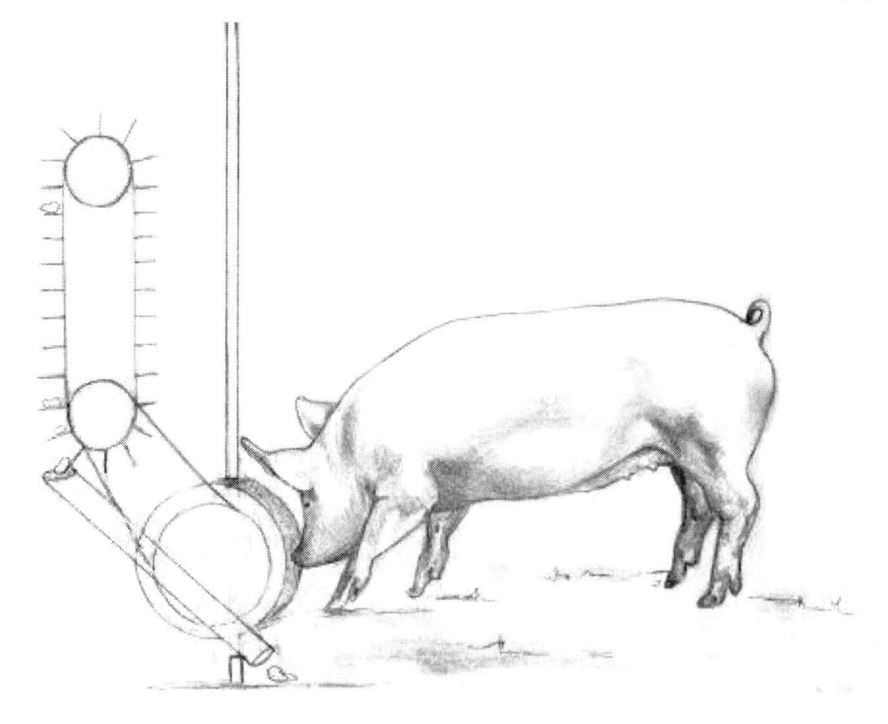

Figure 12-2. Vertical Arrangement of Belt for Perimeter Rooting Device Where Space Outside is Limited

C. Arrange a widely spaced series of ground level sturdy sliding covers that the pigs can root open only with considerable effort. Only when these covers are fully rooted open should there be sufficient space for the pigs to get in their heads to obtain the food treats that you have placed in the pits beneath the covers.

**Recipe 2. Provide Some Puzzles for the Pig to Solve in Finding Hidden Food Treats**

Pigs are very clever at solving puzzles, and you may find it entertaining to devise your own set of challenges for them. Here are some ideas to get you started:

A. Arrange a series of interlocking elements that the pigs must move in some pre-arranged succession in order to obtain food treats.

B. Given just a little motivation, pigs will climb all over apparatus. Provide them with elevated "mazes" that the husbandry staff can rearrange to require different paths to get to food treats.

C. If you have a larger budget, or some special expertise in electronics, a large variety of electronically controlled visual and/or acoustic cues can be used, each requiring that the pig search and root in some different area to discover hidden food treats.

**Recipe 3. Especially Designed for Animals in More Restrictive Environments Such as Those in Biomedical Research Animal Care Facilities**

This recipe is for a device that we originally built for use in a research facility. To successfully follow this recipe you must find a craftsperson capable of precision work with plastic materials. Skilled workers used specifications that I supplied and made a highly reliable device that never jammed and that worked effectively for years. Some less skilled individuals, hoping to cheaply mass produce them so that cost could be reduced to make them more affordable for labs, were unable to produce reliable copies of the device. Careful examination of figures 12-3 and 12-4 will prove helpful in understanding how the successful equipment was manufactured. Access to precision drilling and milling equipment is necessary for making the device durable, long lasting, and reliable.

The device pictured in these illustrations was manufactured by my friends Mark Fribertshausen and Sal Troia who improved on my original design. During their years at San Francisco State University's Science Service Center they collaborated with my students and me in producing many devices that have enriched animals' lives.

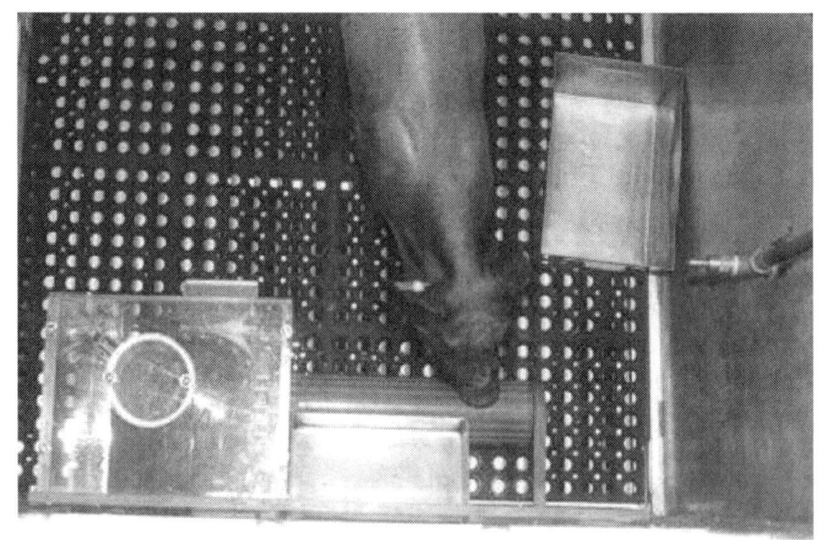

Figure 12-3. Pig Roots on Wheel to Exercise and Deliver Food

1. Build a rotor for the pigs to turn by rooting against it. This can be made of high grade thick wall PVC tubing with strips of PVC welded to it to form ribs. Sharp edges on the ribs must be rounded to prevent pigs from injuring themselves when actively rooting.

2. Install individually removable durable plastic or stainless steel spokes on a circular plate at one end of the rotor. This end must be installed within the protective enclosure that prevents access to the rest of the apparatus by the pigs. This enclosure will house the food delivery part of this mechanical apparatus.

3. Cut a twelve inch diameter, one inch thick rigid plastic disk. If this diameter is impractical for use in your application because of available cage space, make the disk as large as will fit without crowding the cage or preventing easy opening of the door.

4. Decide on the kind of pellets or other dry food that you wish to deliver for the pig's effort. Discern the minimum diameter of a one inch high space that will comfortably hold a small portion of this food. Drill a circular hole of the required diameter approximately one inch in from the edge of the disk.

5. Close to the perimeter of the disk, mount spokes spaced so that they can articulate with the spokes driven directly by the pig's rooting wheel. As shown in figure 12-3, in final assembly the rooting pig will provide the energy to move the disk around in Gattling gun fashion.

6. Install a system for the disk to smoothly rotate by drilling a center hole in the disk just slightly larger than an axle that is made of a rod at least three quarters of an inch in diameter. Connect the axle permanently and rigidly to the bottom of the protective enclosure that houses the feeder and machine a hole in the lid for the top end of the axle to rest in when the lid is closed. If done properly, this will allow you to remove the disk if necessary for cleaning or repair.

7. Choose a convenient place for food to be dispensed from the bottom of the feeder into the pig's habitat. Orient the horizontal disk so that the hole for food collection is above the area of the base that you have selected for food dispensing. Scribe a circle on the base of the enclosure using the hole in the disk as a template. Use a bit just slightly larger in diameter than the scribed circle to make an opening for food to drop when the hole in the disk rotates into position above the opening in the base.

8. Mount a vertical PVC cylinder of approximately 2½ inch diameter to the lid in the manner shown in figure 12-4. This cylinder will be used to hold a supply of the food that will be delivered for the pig's effort. This installation must be precise so that the cylinder will rest on the disk but not impair its rotation. The cylinder should be mounted in a position approximately 180 degrees opposite the food delivery hole in the base of the enclosure. Horizontal alignment of the cylinder must allow the rotating hole to collect food efficiently when the hole passes beneath the cylinder. The cylinder edge must align so that it is within the rotation of the pegs on the disk and above the collection hole in the disk that rotates beneath it. The top of the cylinder should be accessible through a pig proof lid that can be opened to load a supply of food.

9. Install a pawl that limits movement of the wheel to one direction. Otherwise clever pigs will learn to spend only half the rotating effort to gain treats!

10. When completed and assembled properly, the movement of the rooting rotor (no plumbing pun intended) will drive the circular plate. Movement of the plate to the position beneath the cylindrical food reservoir will load the circular opening in the plate with food. Further rooting will eventually drive the plate to a position where it is above the opening in the bottom of the protective enclosure, thus dropping food for consumption near the pig. If the food reservoir is aligned as described above,

centrifugal force will tend to move the stored food toward the position where the circular hole in the disk passes beneath the reservoir.

While the recipe appears quite complex when construction details are described, the principle is actually straightforward and a skilled craftsperson will be able to build a reliable working device. For the apparatus shown in Figures 12-3 and 12-4, we added an inexpensive mechanical stroke counter. This counted revolutions of the feed disk allowing us to monitor use by the pigs.

Figure 12-4. Rooting-Wheel Operated Pig Feeder

Whatever your choice of effective means to enrich their lives, your own life will be richer for having watched the pigs make use of their improved opportunities.

Of course any sensitive pig farmer or zookeeper responsible for the care of pigs will tell you that there are lots of other simple pleasures that one can afford for these wonderful animals. You can also think of some by simply being empathetic. What could be better than a nice cool mud bath in which to wallow on a hot summer day? And maybe while they are relaxing, you could bring them a special *pignic* lunch.....

# 13

# OTTERS

The next few chapters deal with some selected groups of aquatic mammals. I have chosen the species on which to focus based on practicality of enrichment and personal experience. Many of the recipes can be used for a wide variety of aquatic animals. For example, use of the live fish dispenser described in detail in this chapter would be adaptable for enrichment procedures for any kind of marine mammal that regularly consumes fish as part of their diet.

Despite the fact that I have been privileged to study aquatic mammals in captivity and teach about their behavior and physiology for forty years, I confess that my happiest times have been observing these animals in their natural habitats. The most enriching thing that we can do whenever feasible is to leave them there. Of course, there are strong arguments for a contradictory viewpoint. I acknowledge that there would not be the sympathy for marine mammals, and there would be more limited support for laws that specifically protect them, were it not for animals trained in captivity like Shamu and Flipper. We humans have an unfortunate tendency to support and protect species with the best PR and to seldom consider other life forms that are as critical to the balance in nature.

## Some Quick Comments about Enrichment for Otters

There are a dozen or more species of otter, but this chapter will only provide recipes for a couple of the species that some of my graduate students and I have studied both in nature and captivity. I have intentionally selected one otter that is highly specialized for collecting terrestrial as well as aquatic prey and another that is primarily an aquatic feeder. Between these two sets of recipes you can find enrichment ideas applicable for enriching the life of most any otter kept in captivity. Before we go further, I should mention to those of you that feel that it might be better to just provide direct personal affection to these attractive and frisky animals, that otters are not always gentle, sensitive partners in direct interaction. It took a long time to heal the many scars from scratches and bites that occurred during the several months that I worked helping one of my graduate students Pat Foster-Turley with her thesis research at Marine World Africa USA. Pat was an employee of this theme park and she had become custodian for otters confiscated by fish and game wardens from places where they had been illegally kept. While watching the otters scramble and scratch their way up into my lap and occasionally playfully nip my bare legs, Pat gleefully referred to me as "uncle Hal" and applauded the otters. Play behavior that may be fairly comfortable between conspecifics is not always painless when the same behavior is between playful otters and humans. (Yes *of course* it was fun, and I would not mind participating again.)

## Sea Otters

Although no specific recipes will be provided for sea otters (*Enhydra lutris*), I thought it important to mention a couple of incidents that might provide important cautions for those planning enrichment or new naturalistic exhibits for this species. A number of decades have passed since I was invited to attend the much awaited grand openings of two new sea otter exhibits in British Columbia and Seattle. Both institutions had excellent and experienced directors, staff, and planners. They were appropriately proud of the fact that they had incorporated much that was naturalistic in the exhibits. Considerable attention was given to providing stimulation for the feeding behavior of *Enhydra*, including providing small rocks that they could grasp to their chests to use as "anvils" with which to crack open some of their favorite crustacean prey. Both institutions also had fabulous amounts of high quality windows so that visitors could readily observe the beauty of the sea otters' behavior.

Unfortunately, both exhibits had to be shut down shortly after they opened because otters tossed rocks against the windows, causing bad scratches and dangerous fissures. In one case, rocks were dropped hard enough against the bottom of the pool so that it cracked the concrete substrate. If you plan to provide this enriching opportunity for species-typical feeding behavior that is a critical part of the natural repertoire of sea otters, you will need to be cautious in your choice of materials and design of the captive environment.

## Small-Clawed Asian River Otter

Detailed results of some of our research efforts for this species (*Aonyx cinerea*) have been reported earlier (Foster-Turley & Markowitz 1982; Markowitz 1982). The following recipes are derived from enrichment methods that proved measurably successful in those research efforts.

### Recipe 1. *Very* Inexpensive but Labor Intensive

1. Collect some newly emptied coffee cans with plastic lids that snap on tightly. If the otters are able to remove the lids, you will have the additional labor of attaching them with some otter proof fastener each time that the cans are loaded with prey.

2. Punch or drill a hole just large enough for the otters to comfortably reach through in the center of each plastic lid. Deburr the hole to make certain that you leave no hanging remnants of plastic that might injure the otters.

3. Load the can(s) with some appropriate mobile prey such as crickets, and let the otters have fun foraging for the prey and pursuing crickets that escape their grasp. (See figure 13-1)

Figure 13-1. Ultra-Sophisticated Foraging Device for Otters

### Recipe 2. Simple, Inexpensive, Labor Intensive, *and* with the Requirement that You Have Means to Contain Prey that Escape the Otters so that Prey Do Not Infest Your Facility!

1. Work some holes or crevices in which the otters can forage into the existing landscape; or produce foraging areas made of fiberglass or concrete and vermiculite. Be certain that these areas are near some part of the perimeter of the enclosure accessible by husbandry staff.

2. Install a door in the enclosure wall or fence to allow access behind each foraging area.

3. Affix a device such as the screw ring part of the lid for a canning jar concealed inside the area that has the foraging hole or crevice. Orient the ring so that when a jar is screwed on the otters will be able to reach into it.

4. Seal the enclosure area around the screw ring carefully so that prey which are loaded into the jar and placed in the area for foraging cannot escape by any path except the foraging opening.

5. Load the jar with prey such as crickets.

This method has the advantage that it can be made to look very naturalistic and provide a nice demonstration of the foraging abilities of these dexterous otters. If accomplished properly, it should not be terribly inconvenient for husbandry staff to occasionally load some prey throughout the day.

Other advantages include the fact that many kinds of reasonably dry nutritious edibles can be placed in the same foraging area; that at relatively low cost, multiple such enrichment opportunities can be placed in various parts of larger exhibits; and that maintenance costs are typically limited to occasional replacement of corroded screw rings or jars that are no longer sanitizable. Potential problems can be further reduced by having a screw-on top and bottle forged of stainless steel or using a stainless thermos or water bottle for the prey container.

**Recipe 3. For Slightly Bigger Budgets**

1. Decide on a prey species that produces an audible sound, and purchase or build appropriate electromagnetically controlled feeders (figure 13-2) for containment and delivery of prey (e.g., Markowitz, 1982 pp 118-19).

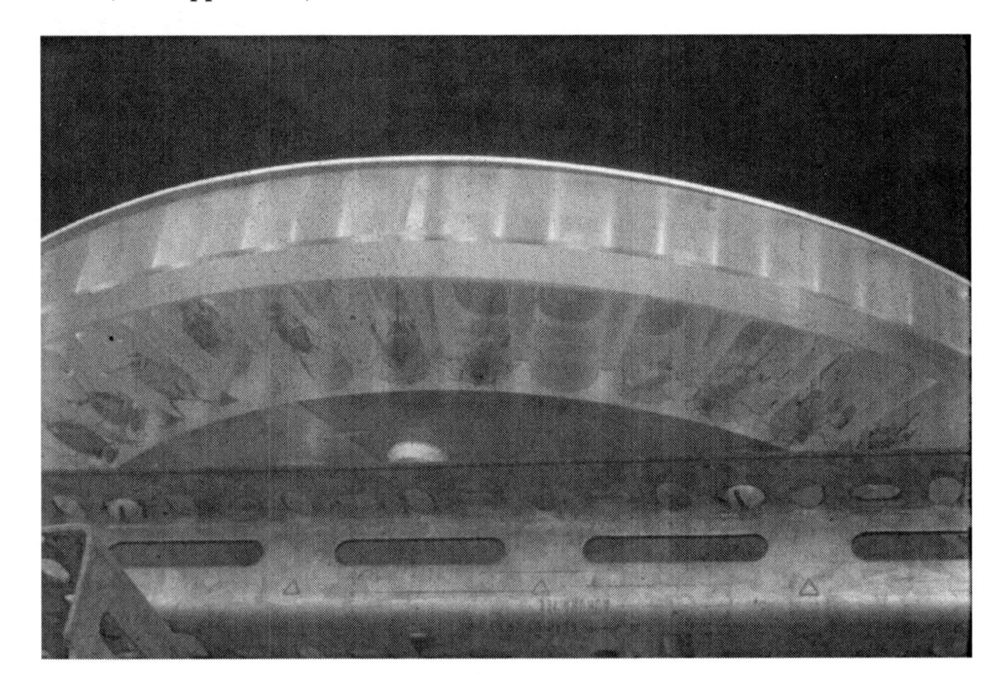

Figure 13-2. Final Residence for Crickets before Release as Prey

2. Follow steps 1 and 2 in recipe 2 above, this time leaving sufficient room so that a feeding device and a waterproof speaker can be mounted and easily accessed in the hollowed out area provided for husbandry purposes.

3. Install a simple detector inside the foraging opening so that it will be triggered when otters put their paws in the opening. (Cheap and reliable options include a properly mounted and matched photo cell and light emitting diode wired to close a circuit; or a passive infrared motion detector.)

4. Record the sounds of the prey species on a chip, and pair the chip with an adjustable amplification stage so that the sound may be adjusted to be audible throughout the enclosure.

5. In a convenient place inaccessible by otters, install a system to control the intervals between times that prey sounds will be heard, and limits the time during which the sound will continue until foraging occurs. Wire this system so that prey will only be delivered when foraging occurs in a period when the electronic prey sound is audible.

A basic flow diagram for the apparatus is presented in figure 13-3.

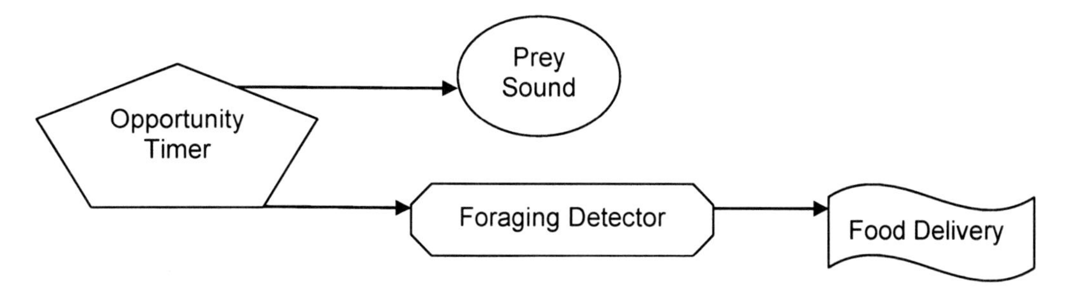


Figure 13-3. Simple Configuration for Using Prey Sounds to Attract Otters to Forage

While more expensive, this system has a number of apparent advantages. The prey can be loaded much less frequently thus conserving husbandry staff time. There will be fewer stray prey wandering around the facility. Consequently, more of the otters' time will be spent in naturalistic hunting behavior because foraging will only be effective when prey are detected to be vocalizing in the area.

**Recipe 4. Still Relatively Simple but Requiring Grander Budgets**

1. Decide how many different foraging areas and kinds of foraging opportunities you wish to provide in a single large captive otter home.

2. Repeat steps 1-4 from Recipe 3 for each of these areas, but this time consider using various appropriate kinds of prey. If, for example, you choose fish as one of the prey species, you can really embellish the design by delivering live fish into a pond or stream when the otters enter the water upon hearing the electronic simulation of splashing fish.

If you use a number of such locations spread around the perimeter of the enclosure, it will be much more cost efficient to use a small computer to produce the sounds and control all other input/output functions. This can be done by running all of the connections from the speakers, prey delivery devices, and foraging detectors through conduit to a central location where the computer can be kept out of danger. In the hands of a competent programmer, a computer costing only a few hundred dollars can easily handle all of the I/O functions for several such hunting areas.

Using a computer for control has many other significant advantages. You can change prey sounds without altering hardware. You can randomize the sequence in which various foraging opportunities occur. You can easily change the duration during which sounds occur and foraging will be rewarded. Perhaps most importantly, you can automatically collect data about when each event including foraging occurs. (Figure 13-4 is a conceptual drawing of how a simple two stage enriched otter habitat of this sort might appear.)

Figure 13-4. More Opportunities to Hunt for Prey Make for Happier Otters

## North American River Otter

First I should mention that there are varied common and scientific names for this species. In doing literature research you will often find it listed as the Canadian river otter and the taxonomy frequently varies between *Lontra Canadensis* and *Lutra canadensis.*

Accomplishing most of their prey capture in the water, these carnivores consume a variety of other animals besides fish, including water voles, muskrats, and aquatic birds such as ducks and coots and their eggs. This means that the enrichment designers need not limit themselves to exclusively using fish as prey. The hunting behavior of this species, and their close relatives the Eurasian river otter (*Lutra lutra*), commonly involves solitary hunting by adult male otters. Females may hunt for extended periods with young offspring, and frequently join with other females or with males during the mating season. Within their territories, otters typically find convenient places to use as "perches" from which they enter the water to capture prey. Repeated use of these areas and defecation there following consumption of fish creates slippery, shiny areas. Experienced trackers often use these "otter steps" to find otters. All three of the following recipes are designed to encourage the species typical behavior of jumping or sliding into the water (see fig 13-5).

Figure 13-5. Otter Water Entry Point

**Recipe 1. Simple and Labor Intensive**

1. Depending on the nature of the enclosure and your available budget, prepare one or more areas for the otters to conveniently slide into a stream or pool.

2. From behind a blind, manually toss one fish at a time into the water when you see that the otter is perched near a slide area. If the otter enters the water and consumes the fish, simply repeat this process whenever the opportunity arises, and as long as the otter persists in leaving the water and returning to this area to watch for and capture fish.

2. Otters that are captive born and have not previously been provided this opportunity may take a while to learn the contingency. You can speed up the process in the following manner:

A. Make raw fish meat balls that are nice and sticky.

B. Observe the otters, and when they approach the area for water entry toss a meatball between them and the sliding area.

C. If they remain some distance from the water's edge, continue to require successively closer movement towards the water until they reach the water's edge.

D. When they are at the water's edge at the top of the sliding area, use your best aim and throw a fish meat ball hard enough so it sticks part way down the sliding area.

E. This will work best if the otter retrieves the treat and continues sliding into the water with it. If the otter begins to retreat to shore rather than sliding the rest of the way into the water, toss a large fish into the water immediately. Depending on the cleverness and skill of both the otter and the human, it may take several repeated efforts employing these steps to encourage the otter to exhibit their natural behavior, and enjoy the chance to "capture prey" in a captive setting.

While this recipe is simple, it requires considerable staff effort. However it is frequently the case that husbandry personnel clearly enjoy educating visitors and interacting with the otters in ways that involve encouraging natural behaviors. The technique also increases vigor in the otters by encouraging frequent activity.

**Recipe 2. Moderately Low Cost, Simple, and Less Labor Intensive**

This recipe will be more attractive for visitors and provide greater variety for the otters if you build more than one area for entry into the water.

1. Arrange a current in the water so that the defrosted fish will be moved along instead of just lying on the bottom. This can be accomplished by using a recirculating pump driven system with a waterfall at one end of an artificial stream.

2. Arrange a simple belt type feeder so that it will deposit fish near the origin of the water turbulence.

2A. For warm climates, convert a horizontal chest type freezer to enclose the feeder belt. The conversion can be accomplished fairly easily by changing the thermostat to one that controls refrigeration rather than freezing temperatures. Bore a hole in one end of the bottom of the chest to accommodate a PVC pipe that will route the fish from the belt into the water. (We built a device like this for dispensing an entire daily ration of refrigerated food for large mammals by using a number of stacked belts and limit switches to transfer successively as each belt's supply ran out. Since it requires a lot of design and construction time and expertise and a greater budget, I would not recommend more than a simple single belt arrangement unless you deem it absolutely necessary.)

3. Focus passive infrared motion detectors narrowly on an area near the water at the bottom of each sliding area that has been created.

4. Convert the output of each motion detector so that instead of turning on flood lamps it can be used to turn on the feeder belt for an appropriate time to deliver one fish. As described in chapter 20, be certain to select a motion detector that allows adjustment of the duration of output signal.

5. Connect the motion detector outputs through conduit to the motor drive of the feeder belt.

With this simple arrangement, sufficient fish can be loaded on the belt to allow husbandry staff to attend to other chores and only reload the belt close to the time all contents have been delivered. If otters occasionally enter the water when no fish are loaded on the feeder belt, this will simulate natural conditions where entry into the water does not always lead to a meal (figure 13-6).

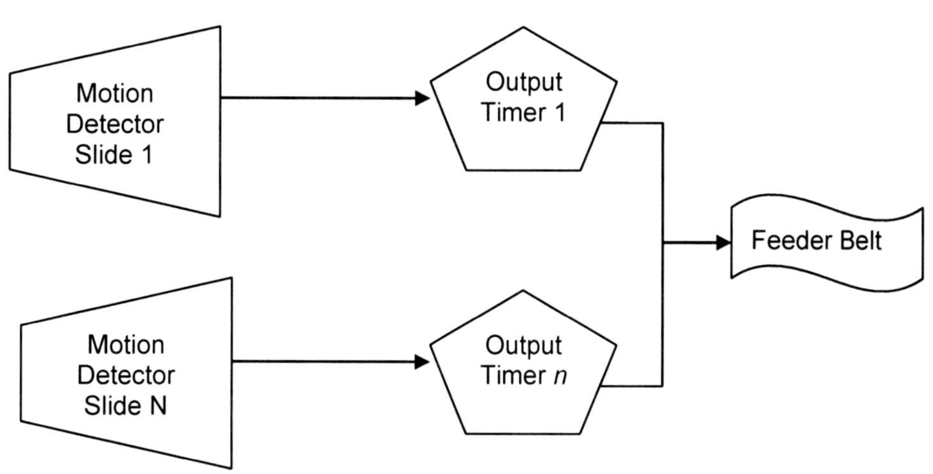


Figure 13-6. Occasional Fish Delivery Upon Using Slide Area to Enter Water

**Recipe 3. More Expensive, More Naturalistic, More Costly Fish, Less Labor Intensive**

1. Arrange two or more attractive water entry areas for the otters.

2. Place a hidden waterproof speaker and a motion detector in a protective mounting near each entry place to the water. Run hidden conduit from this speaker and motion detector to the computer which will control the enrichment equipment.

3. If the enclosure is in a very warm climate, it may be necessary to convert a freezer chest as described in the previous recipe. This time the thermostat should be selected to produce cool (but not cold enough to be immobilizing) temperatures to keep the live fish from overheating while they are in the feeder compartments. In temperate areas it will only be necessary to install the fish feeder in a shady area.

4. The unique component in this recipe is a live fish feeder. This type feeder was invented for our use by my friend Gary Schroeder, a talented craftsman and electronics technician, who used to be employed in one of my funded research projects. It is by far the most reliable type live fish feeder that I have ever seen, and it has the significant humane advantage that it maintains the prey relatively comfortably until they are released into the main body of water. I only present general construction details here, because whoever designs and constructs the device will have their own favorite method for driving the system in a synchronized way to deliver one compartment of live fish at a time into the water.

Where feasible, I would deliver fish from the feeder by means of a sharply downward angled PVC tube into an elevated active water stream or waterfall that will produce random location of the fish as they are launched into the body of water.

Construction details:

A. Procure a considerable continuous length of very strong watertight reinforced flexible tubing with an inside diameter of approximately two inches. Form this tubing into a tight spiral at least ten inches in diameter by winding it around an axle made of heavy gauge PVC pipe or other suitable material. Wind as many segments as you wish to have compartments for fish or as your space allows (see figure 13-7). Cut the spiraled tubing to an appropriate length so that one open end will be oriented with a vertical opening facing upwards, and the other open end faces downwards when the spiral is tightly formed and secured to the axle.

B. Run small individual rigid breather tubes of approximately one fourth inch inside diameter between each adjacent circular chamber that you have formed with your spiral. These tubes will prevent air locks, thus allowing the water to move easily from cell to cell as the spiral is rotated.

C. Mount the axle to a stepping motor or synchronized motor system with limit switches so that the spiral can be rotated *exactly* 360 degrees at each delivery signal. Be certain that the direction of rotation will move the open end that is pointed vertically in a clockwise direction (figure 13-7).

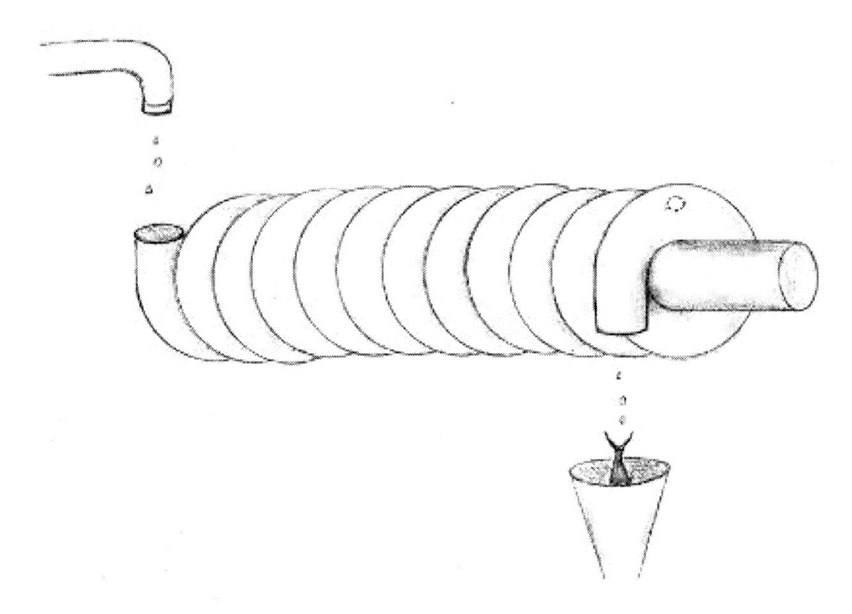

Figure 13-7. A Reliable Humane Live-Fish Dispenser

5. After assembly, test the feeder by turning on a water source at the vertically open end and rotating the spiral one revolution at a time. Water should advance through the cells formed by the spiral as you turn each revolution, and once full the water should flow continuously out of the open end which points downward when at rest. Each cell should have an air space filling most of its top half when at rest and the bottom half should be full of water in this circular "Fish bowl."
6. After testing the operation of the device and making any necessary adjustments, install it so that a water tap can deliver water from a position just above the feeder, thus pouring directly into the other open end of the semicircle when it is in resting position. In order to conserve water, both this input end and the output end of the feeder should feed a common funnel that routes both overflowing water and launched fish into a waterfall that spills into the main body of water.

7. It would be a shame not to collect data once you have gone to the expense of doing all of the above, so use a computer with an appropriate I/O board to record all triggering of the motion detectors, presentations of prey splashing sounds, and deliveries of fish into the water. Program the operation of the enrichment equipment as shown in figure 13-8, with the sequence as follows:

A. Produce the sound of splashing fish near one of the water entry areas at random times and locations. .

B. If the otter is detected entering the water at this noisy area within a pre-selected time limit, deliver a fish through the waterfall and turn off the sound.

C. If no entry occurs within the time limit, turn off the sound and begin another random interval between opportunities.

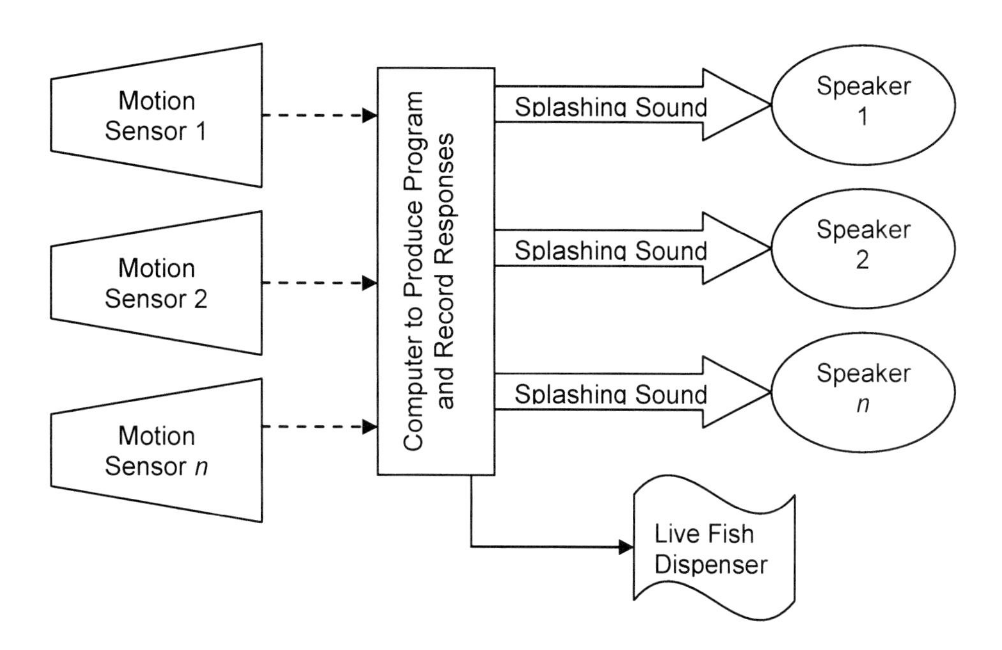


Figure 13-8. Entry Into Water at Splashing Sound Location Delivers Fish

8. There are a lot of "bells and whistles" that most programmers will suggest that you add to this simple program scenario. It has been my experience that the more elements you introduce, the more likely something will go wrong. Yet, there are some attractive options such as adding a counting of spiral rotations to the program so that you can immediately check how many chambers of fish have been delivered and automatically identify when the feeder needs reloading.

While no captive enrichment system can fully capture the true beauty and variety that one may find in natural habitats for otters, this system will produce attractive activity for the resident otters. It also lends itself well to graphics that explain some elements of the natural history of these animals. You can add a large screen monitor and a protective mounting for it for little additional expense. Besides providing a convenient close-up of feeding behavior, the monitor can display video information for the public about the enrichment activity. With modern computers it is a simple matter to store extensive video information and display it along with the view and sounds of live activity captured by a video cam.

Otters are such active creatures when feeding and playing that they offer wonderful opportunities to teach young people about the value of conserving the lives of other creatures besides ourselves. Well designed enrichment that provides increased opportunities for otters to enjoy and control parts of their captive lives can be richly rewarding for all concerned.

# 14

# SMALL CETACEANS

While I anticipate that some readers will be offended by some of the things that I say in this chapter, I can pledge to you that none of it is said with intent to offend anyone. Small cetaceans including the many species of dolphins are truly complex and wonderful animals to observe in nature and to study in captivity. They are not saintly, nor are they universally kind to humans or astoundingly superior in intellect as some have suggested. Like primates, small cetaceans are oriented toward sexual stimulation in virtually every imaginable variety that is anatomically feasible. The inquisitive and often playful nature of cetaceans, coupled with their excellent learning ability, make it possible to suggest an endless array of things that might be used to enrich their lives when they are in captive environments.

This chapter will sporadically include recounting of personal experiences that led to suggestions for enrichment. In addition to recipes of the sort that fill much of this volume, I will take license to make some rather whimsical suggestions that have occurred to me as a function of personal adventures. Most of these are quite different from enrichment ideas that I have proposed elsewhere in the book because, to my knowledge, they have never been tried as forms of behavioral enrichment in the past.

I will also present some recipes primarily derived from my research and collaborative efforts with my students and colleagues. Some of this research investigated the relative value of different kinds of enrichment for dolphins. Other work was focused on specific learning abilities and/or attempted to establish varying kinds of interfaces for human-dolphin communication.

***GOVERNMENTAL WARNING: DO NOT READ THE NEXT SECTION IF YOU THINK IT IS A SIN TO EXPEND SEXUAL ENERGY OUTSIDE OF INTERCOURSE.***

**How About Sex "Toys" for Horny Cetaceans?**

A Few Selected Observations: 1) Anyone who has spent considerable time studying captive dolphins in interaction with humans will tell you that occasionally humans have to ward off sexual advances by obviously aroused animals. 2) Frequently, when new large non food objects are introduced for the first time into their captive quarters, dolphin's investigations include exploration of opportunities for sexual stimulation. For example, the first time that we put underwater hydrophones protected in PVC pipe in an exhibit at Steinhart Aquarium, the female Pacific white-sided dolphins (*Lagenorhynchus obliquidens*) used the poles for genital rubbing. 3) Cetaceans are geniuses when it comes to finding ways to gratify themselves sexually! Many years ago when George Rabb and I were conducting husbandry training sessions in the newly opened Minnesota Zoo, one of the curators gave me a tour of the aquatic facilities. The Beluga whales (*Delphinapterus leucas*) were in an interesting exhibit that was unusually large compared to other captive facilities for this species. This exhibit had a narrow channel that was transparent on the bottom and through which the Belugas occasionally passed. This allowed for relatively close visual inspection by staff and visitors. It also allowed staff members to include in their educational tours a clear description of differences in male and female sexual anatomy. In addition to showing me these exhibit features and telling me about use of training techniques he used to make husbandry tasks involving the Belugas easier and safer, this curator confided that in their attempts at enrichment they had not invented *anything* that seemed as attractive to these cetaceans as the intense water inrush each morning when the pool was filled with additional water. The male and female whales would race to see who could first get their genitalia in the best position to be stimulated by force of this water jet.

All of this is a lead–in to my first whimsical suggestion. If frequency of use by animals and apparent enjoyment of the activity are among our major criteria for successful enrichment, why not provide captive marine mammals with some methods to obtain sexual gratification? Of course one would want to do careful observations to make certain that there were not deleterious effects such as total cessation of breeding behavior or increased aggressiveness toward other marine mammals in the captive environment. There is also the possibility that in some communities the public might be offended by witnessing such behavior. This concern could be overcome by designing some regions of the captive environment that were not available for viewing by visitors, and providing the enrichment there. If the staff was offended, you might consider hiring less sexually repressed people.

I leave the details of species-appropriate stimulation for careful study by anyone who might decide to undertake this radical experiment in environmental enrichment. However, some of the aforementioned experiences might provide clues about starting points.

**Sometimes the Thrill is in the Chase**

When watching small cetaceans provided an occasional treat of live fish for consumption, I have frequently seen them engage in behaviors that allow the fish a head start so that they can chase and capture them. This kind of behavior is not seen very often when observing cetaceans feeding on fish in the sea, probably because in natural circumstances the fish have a much better chance of escaping and feeding attempts are not always successful.

My second untested idea is that some means to simulate a chase would be attractive for those marine mammals that consume actively moving prey rather than plankton. Surely any enrichment design team will enjoy brain storming about means to accomplish this, and I will just toss out one very simple, modest budget suggestion:

**Recipe for Simulating Pursuit of Fish**

1. Where you have a relatively long expanse of wall on one side of the pool, produce an underwater image of fish (e.g. by locating a high intensity projector with a polarizing lens on the other side of the pool).

2. Motorize the projector mount so that it will move the image quickly from one end of the pool to the other with random undulations along the path.

3. Install a motion detector near each end of the side of the pool on which the images are projected.

4. When the cetacean is detected at the starting area for the chase, turn on the moving image.

5. If the marine mammal beats the fish in the race to the other end, as registered by the second motion detector and a device that monitors the position of the motor driven image, automatically deliver some fish.

With a little ingenuity this kind of system could be engineered to provide a variety of "Fish chasing" areas in a captive environment (figure 14-1).

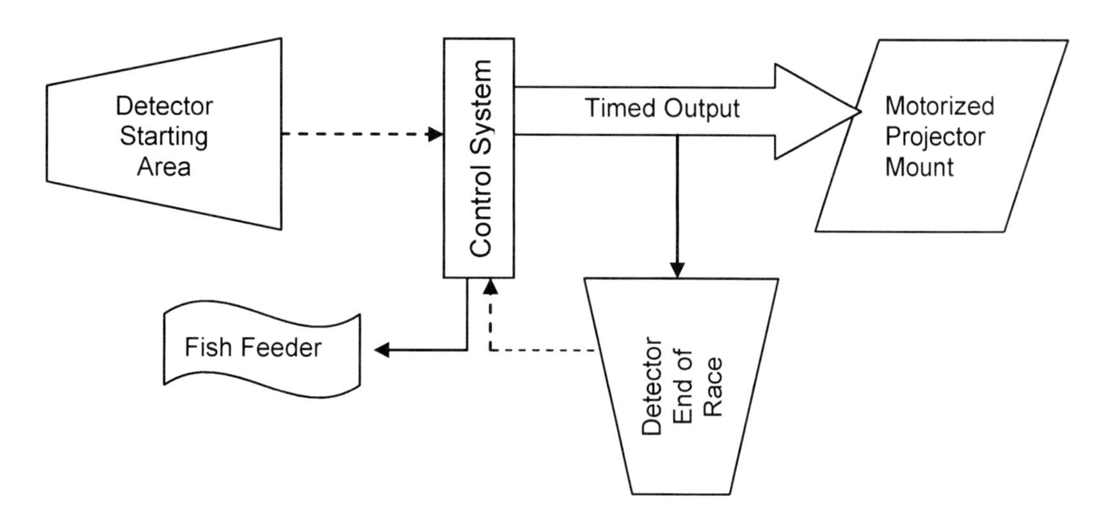


Figure 14-1. Moving Projected Image Fish Pursuit to Capture "Prey"

**A Small Cetacean's Perception of the World is More Highly Acoustical than Our Own Predominantly Visual View**

Incontrovertible evidence of the dolphin's ability to use active echolocation in picturing its underwater environment exists. In formal experiments, it has been demonstrated that this ability allows remarkably fine discrimination of details.

In addition to producing vocalizations in sonar frequency ranges, many small cetaceans also produce a remarkable variety of audible vocalizations, including ones such as "signature whistles" and pod "dialects" that clearly seem to be used for communication purposes. It is remarkable that there is a paucity of published ideas for acoustic based environmental enrichment for captive cetaceans. A recipe for some acoustic enrichment, based on our earlier work, is presented later in this chapter. Here is the 'sketch' of an idea that I do not think has yet been tried:

**Recipe for Acoustic Based Enrichment**

1. Use a motion detector to ensure that there are no small cetaceans in the area of the pool where the activity will be focused.

2. When you detect that there are no cetaceans in that sector, move the typical swimming sound of some prey species through a series of underwater speakers in a succession that ends in an area with at least three visually opaque but acoustically transparent chambers.

3. Once the sound disappears near the chambers, introduce some object such as a small plastic fish into one of the chambers. This can be accomplished by using an inline solenoid or a motor and cam device to introduce and remove the object from the selected chamber.

4. Install your choice of contact detectors on each of the chambers.

5. On occasions when the cetacean goes directly to the box containing the symbolic prey, automatically deliver a fish reward (figure 14-2).

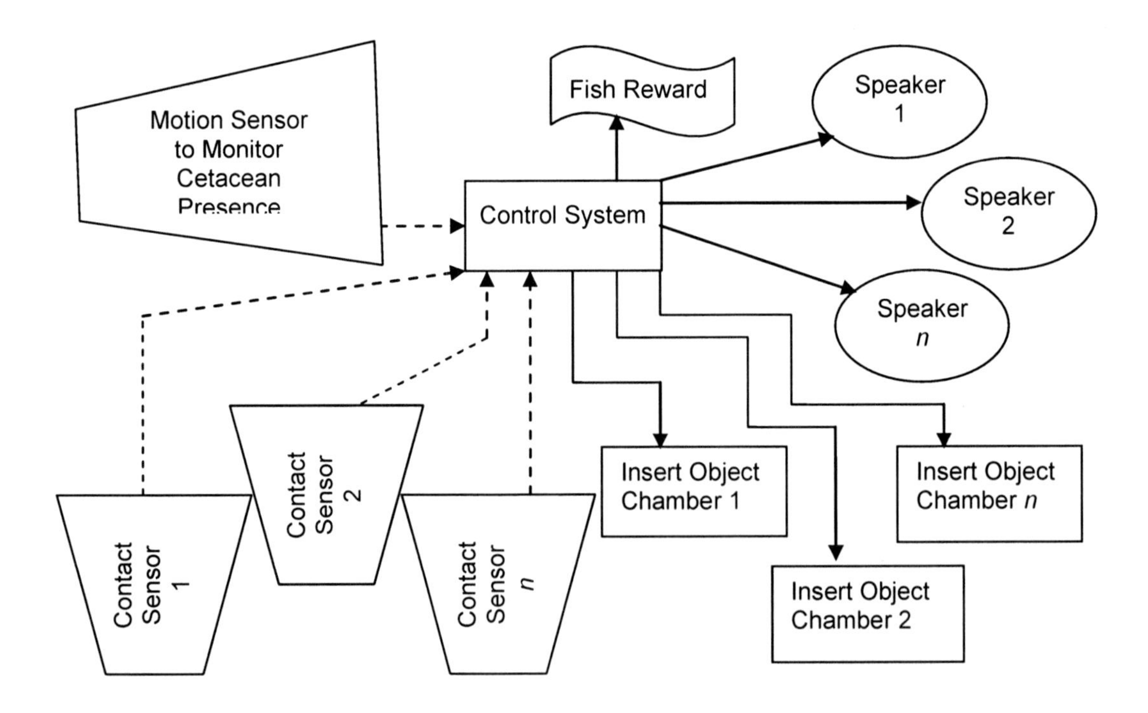


Figure 14-2. Sequential Sound Initiated Search Requiring Echolocation for Obtaining Fish Reward

Although I have never tried this, I have some confidence that dolphins would eventually learn the contingency without any training, and that this would provide some incentive for the active use of echolocation in captivity (figure 14-3).

Figure 14-3. Dolphin Selecting Box with Fish Picture

There are good reasons to do considerable careful observation and simple testing of behavior of the cetaceans in your care before spending money, time, and effort in trying the above recipe. In captivity, the measurable output of sonar range vocalizations by dolphins is much reduced when compared with the prevalence of these acoustical emanations at sea. Some of my students and I, while looking at results of some of our research in this area, have speculated that this may be due in considerable part to the fact that confined pools cause confounding reflections of signals thus "jamming" the sonar sense. On this basis, my first suggestion would be to do the more simple experiments necessary to evaluate whether the dolphins can identify when a visually opaque, acoustically transparent chamber contains an object or is empty. The simplest such experiment would involve just two chambers and the manual loading and unloading of objects while the cetacean is kept at a distance, then delivering rewards on those trials where it chooses the randomly located box containing the object.

I have glibly avoided another problem in my sketch for the apparatus. For this to really require the use of sonar by the cetacean, one has to avoid other sensory cues to identifying the chamber that contains the object. The most obvious of these is sound made by the apparatus that inserts the object into the chamber. Another potential way to identify the loaded chamber is by some visual indicator outside the opaque chamber, such as a distinguishable change in the mechanical "arm" that inserts the object. These are not trivial problems, and it may take substantial preliminary research to ensure that responses cannot be made on the basis of such unintentional cues. For example, even if one employs chambers that are acoustically soundproof to us, these highly acoustically oriented mammals may still detect these sounds. If your intent is to combine enrichment with research, it will be especially important for you to invent a means to obviate these potential ways for cetaceans to use senses other than echolocation.

**Naturalistic Exhibition or Direct Interaction with Humans?**

Many years ago, I introduced a chapter in a handbook of marine mammal medicine by indicating that if I had to choose between a home for a captive marine mammal, and the choices were between a naturalistic museum-aquarium type exhibit and one in which there was considerably more responsive interaction with humans such as a theme park, I would choose the latter (Markowitz 1990). Many of my friends and colleagues told me that they did not at all agree because they felt that training marine mammals to entertain humans in a manner that often made "show biz" characters of them was demeaning. Today, if put in the undesirable position of having to answer this question I would still give the same response. My reasons can be briefly summarized as follows.

First, I know of no naturalistic circumstances for captive marine mammals in which there are appropriate amounts of stimulation for these active and inquisitive animals. Second, the marine mammals most often exhibited in theme parks seem to thrive on learning how to obtain food for their efforts, including working for fish even when they do not choose to consume them immediately. Third, neither the public nor federal regulations would permit some kinds of natural activities, such as killer whales feeding on pinnipeds in captive circumstances.

Husbandry techniques evolved in aquarium facilities that actively train marine mammals are typically safer and more painless for both marine mammals and staff. It is interesting to note that most facilities that maintain marine mammals in more naturalistic appearing zoo and museum aquarium environments now use these training techniques to facilitate health examinations and administer health care. Twenty five years ago it was commonplace to see pool water lowered by dumping hundreds of thousands of gallons of water. Then in the lowered water, small cetaceans, often while suspended in slings, were surrounded by husbandry staff who restrained them when it was absolutely necessary to give them care. Today you are likely to see captive killer whales and dolphins that have been trained to come to veterinary and/or husbandry staff and present their tail stock in order to have blood drawn, to allow closer examination, or to receive inoculations. This not only reduces stress for the cetaceans and staff members, but it increases the feasibility of more frequent health examinations.

Most of the ideas provided for behavioral enrichment in this chapter are based on my assessment that small cetaceans in captivity require much more stimulation for activity than is usually provided them. Stimulation involving important contingencies that appear in nature cannot be provided for these mammals given current budgets for enrichment. Of course, the reader may properly guess that I fantasize about enormous budgets and much larger underwater environments where these wonderful, capable animals would have a great variety of stimulation including species-appropriate ways in which to be able to capture food and to engage in other activities that are attractive for them.

**Some Ideas about More Naturalistic Exhibits for Small Cetaceans**

In many of our field studies of dolphins we have witnessed their entry into mangroves and other visually complex areas in the pursuit of prey (e.g., Grigg and Markowitz 1997). In comparison with the complexity added to aquatic exhibits housing large fish forms (and occasionally to exhibits for otters and sea lions), cetacean exhibits remain relatively barren. Inclusion of an artificial dense mangrove into which live prey were occasionally released might provide both a happier existence for the dolphins and a richer, more rewarding experience for visitors.

A second idea is that where larger exhibits and budgets are available, it would be nice to find a means to incorporate more than one kind of bottom surface material for the dolphins. With a little thoughtful consideration of materials and necessary filtration systems, it might even be possible to include sandy bottom areas that extend outside the mangroves, where dolphins could be observed actively bottom foraging for prey.

I recognize that one of the reasons that dolphins are almost always exhibited in featureless pools is because people enjoy watching the beauty of their swimming behavior. However, providing a place where dolphins could occasionally escape from view and pursue prey seems like a minimum humane enrichment step. Surely visitors would also be entertained by watching and trying to guess where they would next emerge into view. Creative exhibit designers have already invented ways in which visitors cannot visually intrude on the privacy of dolphins, and can still observe them. Unfortunately this does not preclude humans pounding on walls or shouting loudly which clearly affects these acoustically sophisticated mammals.

## Partial Recipes for Untried Ideas

In this section I briefly suggest a few ideas that I have wanted to try but have not yet had opportunity to accomplish. Although untested, all are based on some previous captive studies or field observations.

### Untried Recipe 1. Simple and Expensive

This recipe is intended for classical pools that are somewhat rectangular in shape.

1. Get an "angel" to buy or donate the best available wave-making equipment.

2. Install a simple means, such as waterproof transluminable touch controls, for the mammals to use in turning on the waves.

3. Adjust the wave machine so that the waves always begin slowly and build to the maximum size selected by the marine mammals.

Having spent hundreds of hours watching inshore dolphins "body surf," I have confidence that they would enjoy the opportunity to exercise in more turbulent waters. This kind of apparatus would also be useful for a number of pinniped species. For example, I have watched Steller sea lions (*Eumatopias jubatus*) body surfing for more than an hour at a time off the Oregon coast. Close observation with a high power scope revealed that there was no foraging behavior involved and no apparent social purpose for this behavior. Like the dolphins, the sea lions simply went out and timed the crest of a wave to surf in, "bailed out" before reaching shore, then returned to wait for another large wave.

Figure 14-4. A Surfing We Will Go!

**Untried Recipe 2. Complex and Expensive**

This recipe is intended for research/enrichment in which the dolphins learn a vocabulary and some syntax for exchanging information with humans. (It could easily be modified for use with other kinds of animals such as pinnipeds and apes.)

1. Arrange a series of six frosted glass screens with rear projection equipment allowing a variety of images to be shown on each screen. The dimensions are not too critical, but the screens should be at least four inch squares and no larger than eight inch squares. They should be placed in a single horizontal row and spaced about two inches apart.

2. Make an identical series of equipment, and install it in a row about four inches below the first row, making certain that the screens are in line horizontally so that there are two screens in each vertical row. The bottom of this row should be just slightly above the surface of the water unless you have truly waterproof equipment in which case the installation may be underwater. (Research in which I have been involved has shown that dolphins are capable of transferring information learned above the surface to underwater identification of the same symbols.) Either make each screen in this bottom row touch-sensitive, or arrange some optical or mechanical means so that when the dolphin touches the screen a response is transmitted to the computer.

3. Install a feeding device such as a live fish feeder or a belt type feeder in a location where food will be delivered into the water near the screens.

3. Arrange appropriate I/O connections between the peripheral equipment and a computer.

4. Employ an excellent computer programmer (unless you are one).

5. Produce programs to progress through the following sequence of steps as the marine mammals master each requirement. (If properly accomplished, this equipment could serve for many years of interactive enrichment and communication research.)

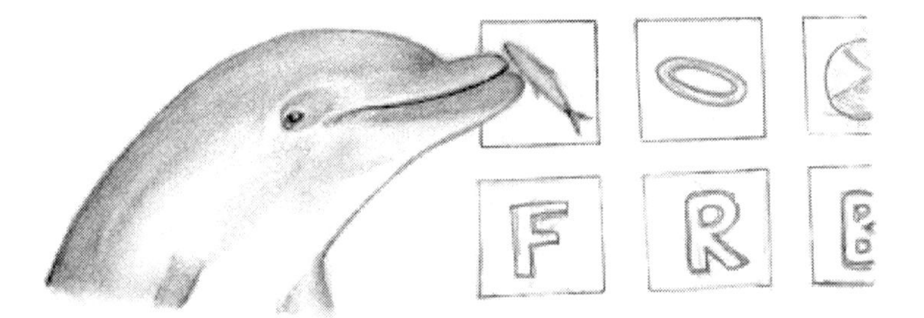


Figure 14-5. Dolphin Vocabulary Training

Program A: The top screens randomly display six pictures (still or motion). The screen below each of the pictures displays the correct symbol for the objects or action shown in the top picture. Touching any of the screens in the bottom row leads to delivery of fish. This is followed by a thirty second interval before the next random display.

This first program is for use until the dolphins have been exposed to the stimuli for a number of sessions. It will also come into use for "remedial tutoring" when correct symbols are not employed in later stages of the research.

Program B: The bottom row randomly displays the six symbols. The top row displays only one picture *in a random location.* The dolphin is allowed to continue trying until it correctly touches the symbol for the picture shown in the top row. When the correct response is made, fish are delivered and a thirty second interval ensues before the next random presentation. Once two consecutive errorless responses are made by the dolphin, the program changes to one that rewards correct responses and produces a fifteen second interval without projected images whenever an incorrect response is made.

Inventive researchers and enrichment technicians will undoubtedly want to add embellishments to this relatively straightforward recipe. For example, if the top row also includes touch sensitive screens, you could program randomly arranged pictures of actions or objects in one row and randomly arrange the symbols for these objects or actions in the other. Dolphins could then learn to touch the picture of the action or object and then touch the matching vocabulary symbol. Once these skills are mastered, it would be logical to add syntactical requirements, such as requiring that the symbols for fish and movement through the water both be selected when the picture is of a fish swimming.

## Recipes Comprised of Components that have Proven Successful

### Recipe 1. Detection of Acoustical Reproduction of Sound by Dolphins

This idea is largely based on some research equipment and procedures that Doug Richards showed me a number of decades ago in the Koala Basin Dolphin Research Facility of the University of Hawaii.

Since dolphins are acoustically sophisticated animals, it was not surprising to learn that with patience one could teach them to reproduce complex contours of sound that were played to them underwater. Doug and his colleagues discovered that dolphins could easily respond by mimicking the "melodies" of brief underwater sounds in order to obtain reinforcers as long as they widened the acceptable envelope of pitch for the contours in the sound that they were required to reproduce. What *was* surprising was that when they sang back these brief "refrains," they did not necessarily do it on key.

Complete description of equipment necessary is beyond the scope of this book. It requires considerable time and effort to produce a system following the outlined steps below, but the effort would be rewarded with both opportunity for research and significantly heightened acoustic activity by the dolphins.

Step 1. Befriend and/or employ a talented acoustic specialist to help you in the design and selection of components.

Step 2. Install a high quality underwater speaker system to deliver sounds to the dolphins.

Step 3. Install a high quality hydrophone to receive the acoustic responses of the dolphins.

Step 4. Connect this hydrophone to an acoustic analysis system that has sufficient sophistication to differentiate the brief "songs" that you wish to identify and to allow you to see whether these are matched by the second acoustic input (the dolphin's responding vocalization). The equipment must also allow you to adjust the required precision of the reproduced sound (e.g., how far can it vary from the pitch of the sound to be mimicked).

Step 5. Use a computer (which will probably already be required for your sound analysis equipment) and program it to deliver fish as rewards for accurate dolphin reproduction of brief whistled songs (or other acoustic materials that you wish to use to test their ability to correctly reproduce).

**Recipe 2. Simple Apparatus for Cetacean Selection of Enrichment**

This apparatus is based on some enrichment work and associated research that we conducted for a number of years at the Steinhart Aquarium of the California Academy of Sciences (see fig 14-6). The design and choice of components were refined by trial and error in that work. Pacific white-sided dolphins (*Lagenorhynchus obliquidens)* and harbor seals (*Phoca vitulina*) were the marine mammals involved in these efforts. The elements of this recipe, when properly constructed, should be both reliable and durable in an aquatic environment. Total cost depends on the kinds of enrichment offered and the extent to which this equipment is automated or enrichment is manually provided.

1. Make a series of vertical "keys" of successively increasing lengths out of heavy gauge three inch PVC tubing. (The number of keys is arbitrary, but the more that are used the greater the variety of options that can be offered.)

2. Arrange these PVC keys in a xylophone-like array by drilling carefully aligned holes about three inches from the top of each key and placing them on a Delrin axle. Mount the keyboard sturdily to a housing that is appropriate for attachment somewhere on the aquarium perimeter. This housing must be mounted so that the keys will all be partially submerged. Spacing between the keys is necessary to ensure reliable settling of the keys to the original position following their operation. This is most easily accomplished by cutting appropriately thick pieces of Delrin and drilling holes in them so that they can serve as washers to space the keys and to keep the keys on each end of the apparatus from sticking on the axle mounting.

3. In the top end of each key where it protrudes above the edge of the pool, install the sealed permanent magnet portion of an inexpensive magnetically operated switch (the kind that is ordinarily used to identify the opening of windows or doors). Note that this part of the switch mechanism does not require any leads since it employs a permanent magnet. This is important because it eliminates worries about any connections deteriorating with repeated movement of electrical wires.

4. Make a strut that runs horizontally from end to end of the apparatus just above the keys, and attach the encapsulated reed relay portion of the magnetic switches to this strut. Before attaching the relay elements permanently, use an ohm meter to make certain that you have placed them properly. When the keys are hanging directly vertically and motionless, each switch should be operated by the associated permanent magnet.

5. Place heat-shrink insulation around the wires that come from the reed switches, and connect them to wires leading to a control apparatus. If the apparatus is regularly splashed or cleaned with water sprays, it will be necessary both to apply heat to tightly shrink the insulation around the wires, and to then coat this part of the apparatus with sufficient epoxy to make it waterproof. In most installations, you will find that using multiconductor color coded cable that runs through protective conduit will be the easiest way to route the leads from the reed switches to the control apparatus.

6. Carefully decide what choices of enrichment you desire to make available for the dolphins to select. Below are just a few of the items that we have found are attractive to dolphins in this kind of xylophone operated enrichment "cafeteria," and that can fairly easily be automated so that constant human presence at poolside is not required.

A. Reasonably high pressure streams of water delivered at an angle into the pool by means of a solenoid valve wired to the control apparatus.

B. Sounds delivered simultaneously through underwater and in air speakers. You can use various keys to allow the dolphins to select sounds that they wish most to hear. (We were surprised to see that they swam in different rhythms and patterns when they ordered Mozart than when they ordered Carly Simon or James Taylor!) It is also interesting to study the responses to recorded sounds of conspecifics if you use a key to make that an available choice for the dolphins.

C. One of the options that is most likely to be quickly mastered by the dolphins is the ordering of fish (delivered from an automatic dispenser as earlier described). However, this is such a powerful reinforcer that if it is always available as an option the proportion of selection of other kinds of activities may be greatly reduced. Like most animals that we have studied, given the opportunity to collect prey carnivorous marine mammals will persist in the behavior long after they are satiated.

Figure 14-6. Dolphins Place Order for Enrichment

If you wish to have sessions in which humans are directly interacting participants in some of the enrichment options, experience has shown that dolphins will "order" the following things *assuming that the human is someone with whom they wish to interact*. This is an especially serious matter if you are working with species such as white sided dolphins since these animals are very selective about with whom they wish contact.

A. Human tactile stimulation.

B. Entrance of familiar humans into the tank to swim with the dolphins. It is important that the humans have good swimming skills and are aware that interactions of this sort with dolphins involve some inherent risks.

C. Retrievable toys such as large floating rubber balls or circular rings that they can move around with their rostrums. (This may be rather labor intensive for the humans who retrieve the toys!)

7. Connect the peripheral equipment to a reliable low cost computer with an appropriate I/O board as an interface.

8. Produce a program that allows each xylophone key to activate a different type of enrichment. Also have the program provide the following necessary elements for this xylophone to work efficiently as a selection device.

A. "Debouncing" of signals from the keys as they swing back and forth before returning to rest. This can easily be done by causing outputs for signals from the key switch closure to only be registered when they are separated by an interval ample for the keys to be at rest following responses.

B. Allow enrichment staff to adjust the duration of each kind of enrichment that is offered for a single key operation. (Note that the dolphins can be allowed to extend an activity by making additional responses on the appropriate key each time that the key comes to rest.)

Although this recipe suggests the use of a computer as a controller, an experienced electronics technician can also design and build rather simple hard-wired equipment to control this kind of enrichment scheme. However, there are great advantages to using a computer to both provide control of the program and to simultaneously allow time-based registry of responses by the dolphins. A good programmer can include the elements necessary for immediate readout and charting of the current progress of the dolphins in learning to use the xylophone. This data can also allow enrichment specialists to adjust available enrichment types so that the ones seen to be most attractive to the dolphins can be frequently included as selectable options.

**Recipe 3. A Simple Less Expensive "Modular" System Allowing the Cetacean to Order Enrichment**

1. (Optional) Install an underwater speaker.

2. Install a simple device for the dolphin to use in ordering enrichment (e.g., an underwater contact detector or a narrowly focused motion detector).

3. Hard wire the dolphins' ordering device to operate some enrichment apparatus such as those listed in step 6 of the recipe immediately above.

4. (Optional) Make a simple control apparatus that at random times delivers a distinctive sound through the underwater speaker, and responses to the sound will lead to delivery of enrichment.

This approach has some advantages in that one can start with a relatively small budget and add components as additional funds become available. It also means that at times when inevitable service is needed for some components, this does not require the shutting down of all aspects of a more complex system.

I wish those of you who undertake any of these or any other kinds of efforts in enriching the lives of captive cetaceans enormous success. Few kinds of captive animals are in such obvious need of more responsive environments.

# 15

# SEALS AND SEA LIONS

Although there are 36 species of pinnipeds, many of which have subspecies that exhibit distinctive behavioral traits, this chapter derives most of its recipes from my more extensive experience with harbor seals (*Phoca vitulina*) and California sea lions (*Zalophus californianus*). Fortunately, with some minor modifications, the same recipes can be used to provide opportunities for any captive species of pinniped. The previous chapter has a number of recipes that would also serve well for pinnipeds. I will begin the recipes in this chapter with some suggestions for modifications to make some of these better suited for seals and sea lions.

While the majority of pinnipeds have fish as the major staple in their diet, regionally and seasonally they feed in an opportunistic manner on a number of other kinds of prey such as squid, octopus, whelk, shrimp and amphipods. In the wild, there have also been observations of seals lucking on chances to capture and feast on birds. This knowledge provides the opportunity to use a more varied diet as a form of enrichment and stimulant to investigatory behavior by seals and sea lions. However, it is critically important that one study the feeding behavior of pinnipeds in detail before embarking on enrichment for these mammals. Here are a few examples of reasons for this need:

1. In nature there are great differences in periods of fasting for various species depending on many aspects of their natural history and seasonal changes in local resources. The most famous of these exist in Northern elephant seals (*Mirounga angustirostris),* where males seasonally fast for two to three months, and females nursing young transmit astonishing amounts of nutrition to their offspring while simultaneously fasting for four to five weeks. A number of pinniped species are also seen to show fasting at the time of molting.

2. Our research groups have observed and recorded some rather astonishing neophobias in captive harbor seals. These not only involve reluctance to feed on new kinds of prey, but reluctance depending on the orientation of prey fed to them as well. In one such study, we found that seals that were always hand-fed fish presented head first would reject fish presented tail first. Similarly seals that

had always been fed cut up fish would not readily consume whole fish, and those initially fed whole fish would reject fish parts.

3. I have witnessed cases where concerned veterinary or husbandry staffs have resorted to extreme measures such as intravenous feeding for otherwise healthy seals that simply ceased to eat for a while. When I was asked for advice and collected necessary information, I almost always found that the length of the period that they had failed to eat was well within the range of fasting durations for the same species in the wild.

4. Because the vitamin content in fish degrades even when food is stored in appropriate sub-zero freezers, many facilities prefer to store frozen fish only for short periods. They purchase appropriate fish that are the most seasonally abundant for economic reasons. When herring became seasonally available, we always encouraged the switch to this more highly caloric resource from the smelt that were the typical diet of seals in many captive facilities. Predictably there was feedback from some facilities indicating that many of the seals who had always fed on smelt would not eat the herring. There were no pinniped fatalities reported from those facilities that followed the advice to "simply keep offering them herring until they get hungry enough to eat it."

My untested pet theory is that the tendency to avoid unfamiliar kinds of food may ultimately result from selection pressures in the milieu in which some pinniped species and their immediate ancestors have tended to forage. In frequently murky inshore areas, some of the foraging may actually involve use of the vibrissae to tactually locate prey. Some of the items encountered and potentially consumed in these environments may be unhealthy or, in the worst case, poisonous. Consequently, young pinnipeds that often accompany their mothers in early foraging efforts may benefit from the tendency to consume safe familiar foods. In this manner, an evolved heritable tendency toward neophobias in feeding may be beneficial for seals showing this trait. Whether this theory holds up to scrutiny or doesn't, I am certain that gaining a real familiarity with the feeding behavior of a species before planning enrichment involving food consumption is of critical importance in the success or failure of enrichment efforts.

**Modifying the Aquatic Prey Chase for Pinnipeds**

Use the same recipe provided for small cetaceans with the following modifications:

1. Project the image of the prey so that it descends from the surface, requiring the pinniped to pursue it as deeply as the pool allows.

2. Check to see that your end of the chase motion detector will work efficiently at this underwater level. If not, install a different means to detect time of arrival of the pinniped at this area, such as a photic device triggered by interrupting the light when the seal or sea lion passes.

**If You Can Afford a Wave Machine, Sea Lions Will Enjoy the Opportunity to Surf**

It might even be possible with a long enough pool to provide an active environment for compatible small cetaceans and pinnipeds, and to at least occasionally introduce some live prey for them to pursue in this wave filled pool.

**Using the Discrimination Apparatus Proposed for Cetaceans**

Option 1: Arrange a means for the same discrimination apparatus to be used by pinnipeds and small cetaceans by having a gated pool arrangement that allows control of access to the area.

Option 2: Observe what happens if you allow a mixed group of cetaceans and pinnipeds in a larger pool to opportunistically use the device as they wish. (It might really be interesting to see if they interfere with each other or cooperate in solving discrimination problems.)

Pinnipeds have been shown to be adept at solving discrimination problems and learning how to follow simple syntax rules when two or more symbols or gestures are presented sequentially.

**Pinniped Use of the Underwater "Xylophone"**

In our research efforts, the harbor seals actually began to operate the xylophone-like keys before the dolphins in the same pool. We guessed that this might be true because they were accustomed to hauling out on the edge of the pool and the xylophone apparatus was attached to the wall on which they often rested when out of the water. (See figure 15-1)

Figure 15-1. Harbor Seal Responds on Xylophone While Dolphin-Scientist Observes

If you have compatible species and the pool is large enough, simultaneously providing more than one species of marine mammal the opportunity to order various forms of enrichment can be great fun.

## Other Recipes for Pinniped Enrichment

**Recipe 1. Simple and Fairly Inexpensive**

1. Install a number of small waterproof speakers just above the surface of the pool in as widely spaced locations as possible. At each speaker location, install a passive infrared motion detector focused on the area of the pool just beneath the speaker.

2. Depending on what you have available, either use equipment capable of wirelessly broadcasting sounds to individual speakers, or hard wire these speakers through conduit to the control area.

3. Install a means to deliver fish to random locations in the pool. This can be accomplished in a number of ways such as delivering the fish into a waterfall, arranging a motorized slingshot or skeet shooter that moves back and forth in a semicircle slinging fish into the pool at unpredictable locations, or simply introducing live fast swimming fish.

4. Use a simple computer or other controller (Figure 15-2) to operate the following procedures:

> A. At random times and random speaker locations, produce the sound of splashing fish for a preselected duration. The control program must preclude speaker locations where a pinniped is currently detected by the associated motion sensor.
>
> B. If a pinniped is detected swimming into the area while the splashing sound is still on, turn off the sound and simultaneously deliver fish into some random area of the pool.
>
> C. Once the sound ends, whether fish have been delivered or not, generate another random time interval before the next opportunity.

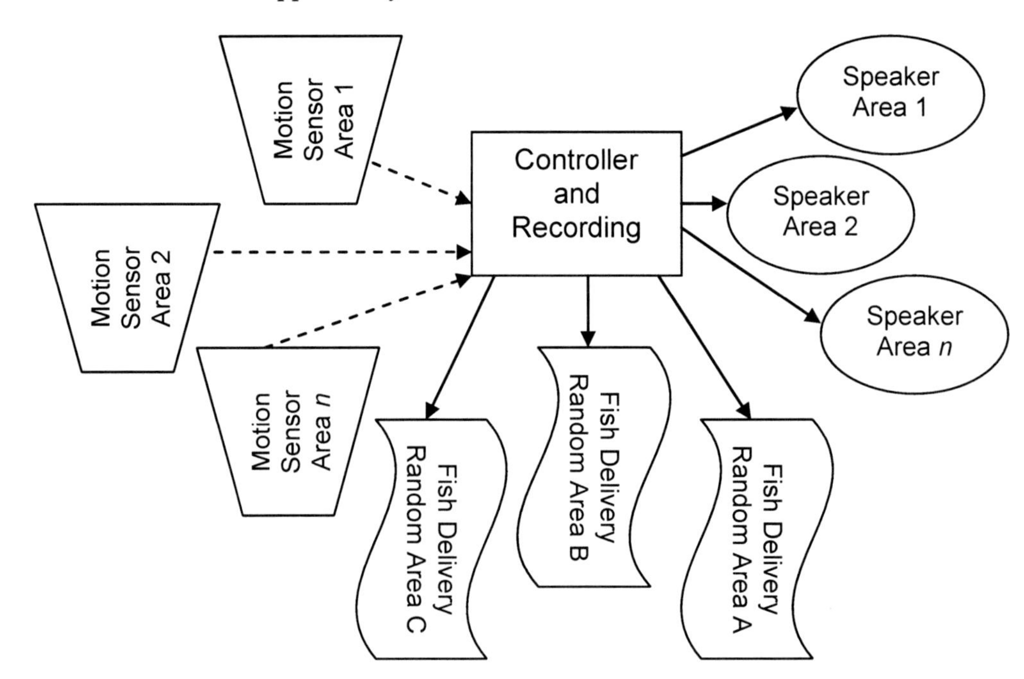


Figure 15-2. Arrangement of Simple Enrichment for Pinnipeds

This apparatus should generate some healthful activity for seals or sea lions. The only caveat based on our earlier studies is that some pinnipeds may learn to exploit others by letting them race to the splashing sound and then competing with them for the fish. Random distribution of the fish helps to reduce this probability, but in the long run the seals or sea lions will sort things out themselves. Interesting and potentially publishable results can be accumulated by having trained observers collect data concerning the seals' trends in exploitation and resistance to being exploited.

**Recipe 2. More Complex**

Pinnipeds are known to differentially identify the sounds of offspring and to ward off the feeding entreaties of what my friend Burney LeBoeuf has labeled "double mother suckers." It might be interesting to provide an enrichment opportunity which would simultaneously allow researchers the chance to assess some of the acoustic discrimination capabilities of the captive pinnipeds.

1. Install a high quality speaker either at the surface or underwater.

2. Install a system for the delivery of fish into the pool.

3. Install some means for the pinniped to respond to some sounds and avoid making responses to others. Although simple manipulanda of any sort can be employed, I suggest a voice triggered apparatus so that the pinniped can vocalize in response to some sounds and abstain from responding to others.

4. Record a variety of appropriate vocalizations such as the sounds of young conspecifics.

5. Program an inexpensive computer to allow you to accomplish both the training and final aspects of this enrichment activity. Although individual enrichment designers may well choose other systematic programs, here is a simple one (figure 15-3) to consider:

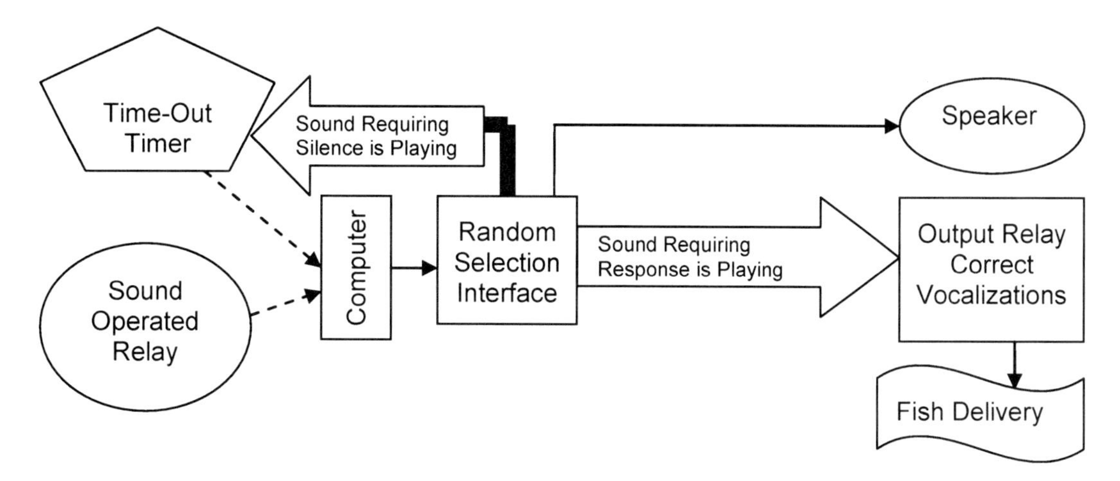


Figure 15-3. Suggested Program for More Complex Pinniped Enrichment

A. Select a number of sounds to which you wish to teach the pinnipeds to actively respond. Also select an equal number of sounds to which you will teach the pinnipeds to restrain from responding.

B. Teach the seals or sea lions how to get payoffs by gradually increasing the number of sounds of each kind. Do this by randomly selecting one of the sounds that are intended to cue responses and one to cue non-responding. Then reward with fish for appropriate behavior. Depending on the pinnipeds involved in this procedure, you may have to produce lengthy "time out" intervals for responding in the presence of the wrong sounds. This is likely to be necessary to prevent seals from responding to every sound presented and taking the fish that are delivered 50% of the time rather than spending the effort to learn the discrimination task.

C. After a number of these differential kinds of responses have been learned to a satisfactory criterion such as ten in a row correct, progress to the final program described in step D below. When you want to add additional sounds, it will be necessary to return to these earlier teaching procedures.

D. Design a program to deliver randomly selected sounds at unpredictable times for pre-selected durations. If the pinniped restrains from responding to an "incorrect" sound ***or*** elects to respond to a "correct" one while the sound is still ensuing, deliver a fish.

**An Alternative Use of the Apparatus Just Described**

Some of my graduate students have done studies of the distinguishing characteristics of pups' vocalizations that may provide information which harbor seal mothers use to identify their own offspring in crowded milieus (e.g. Khan et al, 2005). An apparatus such as the one detailed above could also be used to test what aspects of the vocalization are necessary for differential responding. This might be a neat way to combine enrichment efforts with research and potentially attract extramural funding.

**Some Simple Suggestions for General Improvement of Captive Pinniped Environments and Feeding**

In some happy moments in the past, I have had great fun allowing free-ranging sub-adult male California sea lions to "pursue" me while I snorkeled in areas near islands in the Sea of Cortez. They would typically race up to try and surprise me and then noisily swerve off at the last moment. Where underwater caves were available for these sneak attacks, this became especially interesting for me, and apparently more fun for the young pinnipeds. As mentioned earlier, I have always grieved that so few captive facilities have any true underwater complexity. For sea lions, it would seem especially important and not excessively difficult to provide some caves and other underwater openings from which they could emerge, and into which they could disappear from sight.

Another fairly inexpensive and potentially enriching idea would be to randomly introduce a much wider variety of food types than are typically fed to pinnipeds in captivity. For many pinniped species, this might include more than one kind of live fish and some cephalopods or shrimps. For animals destined for reintroduction, this procedure might much better prepare them for the challenges of competing for resources when released in the sea.

In closing, for those directing procedures in rehabilitation facilities, I would especially suggest that you might wish to employ automatic live fish feeders whenever possible to reduce the learned dependence on humans for sustenance. Providing methods for pinnipeds to learn to pursue prey in captivity without the necessity for constant human presence will certainly enhance their fitness for reintroduction to the sea.

# 16

# WOLVES AND OTHER CANIDS

Perhaps the most wonderful thing about being a teacher is that you learn so much from your students. Many students have become part of our extended family. About a third of a century ago I had a student named Paul Paquet, who had spent some time as a naturalist. Paul taught me some things about otters and much more about wolves and coyotes. Many colleagues have extended my education about canids. My friends John Fentress and John Sullivan, who loved and lived with wolves and coyotes in and near their homes, were especially generous about helping me to learn more.

As these friends and other kind experts taught me more about wild canids, I began to more fully appreciate the nature of pack life, and how important it was for these animals to merge themselves into a community of life. Odd as it may seem, when I was in John Fentress's home in Nova Scotia or John Sullivan's place in southern Oregon, I always felt like part of a pack because they introduced me to their canid friends who lived with or near them.

This got me to thinking about how strange and limiting life in institutional captivity must be for canids, who most often are deprived of a chance to identify and defend a home and to choose companions. Here are a few enrichment ideas, some that we have tried and some that I have dreamed about over subsequent years. Although I will specify wolves in these examples, most recipes would work for many non-domesticated canid species.

**Acoustic Enrichment**

Everyone knows about the exceptional ultrasonic hearing of dogs and also recognizes that calls are an important constituent of life for wild canids. So it seems only natural that we should look toward sound as a means for stimulating behavior in captive wolves. Before introducing a very simple and inexpensive method that proved useful in generating pack behavior in captive wolves in the Portland Zoo, I should confess that we came upon a critical part of the recipe by blundering!

Since all of us liked music and had high fidelity equipment to ensure that sounds were faithfully reproduced from recordings, we figured that acoustically able animals like wolves would be most responsive if we delivered them very high fidelity sounds. I made some fine recordings of the calls of wolf packs other than the one that we tended in captivity. Then I arranged to place speakers in places where these sounds could radiate into the grotto without anything to distort or baffle them.

For reasons that those of you who have worked in facilities dependent on public monies for survival will understand, we were asked at the last minute to demonstrate our experiment to visiting dignitaries during a brief interval in their tour of our facility. My helpers rushed around to gather the needed equipment from our lab and bring it to the grotto. In our hurry, we forgot to bring extension cords and there was no electrical outlet near the area where the sound generating equipment was to be used. Under pressure from the tour guides to start immediately, I reluctantly decided to "give it a try" by plugging in the system far down a corridor and cranking up the volume. To everyone's surprise, upon hearing the sounds, lazing members of the captive pack jumped up and exhibited all sorts of wonderful affiliative behavior. They nipped gently at the neck of the dominant male, began to echo back calls of their own, and stayed active and apparently interested long after the recorded sound ceased being reproduced.

Later, purists that we were, we tried the system as originally planned. The sounds were played directly to the wolves with high fidelity rather than through the labyrinth where they had originally been played. To our dismay, the wolves showed very limited response to the recordings. At first I thought that they must have quickly habituated to the sounds, but this did not exactly make sense to me since we were using new sounds recorded from a pack they had never before heard. Then it struck me that we should go back to the method that had been successful and try playing the sounds from the more distant location in which there were acoustic impediments before these sounds reached the wolves in the grotto. This time the acoustic stimulation worked effectively, and the wolves got quite excited. After the fact, it is easy to make up plausible explanations for the importance of having the sound reach the wolves through impediments. My favorite ad hoc guess is that in the wild wolves hear other packs at a distance, and often through impediments like trees. So, playing calls from a distance and behind barriers may make them sound more genuine to the wolves. But, of course this is only a WAG (wild ass guess). What matters for enrichment purposes is that repeated tries of this in other facilities have shown that recorded calls played at a distance and through some barriers generates considerable activity and interest for captive wolves.

**Recipe for Simple Inexpensive Acoustic Stimulation**

1. Use a high quality recorder to record the sounds of a number of wolfpacks. If possible, collect some of these in nature, and supplement them with ones from captivity. Alternatively, you can obtain good recorded wolf calls from one of the libraries of animal sounds.

2. Select a location to reproduce the sounds, taking care to avoid areas where these calls might have deleterious effects on other captive residents of your facility.

3. At infrequent intervals such as every few hours, try playing the sounds and carefully document any changes in behavior of the wolves. Adjust the location, intensity, and sound baffles to find what works best.

In my experience with this technique, there have been no damaging outcomes for the wolves in our care. However, there are critical times when it might be unwise to provide additional stimulation such as that described in this recipe. For example, when dominance orders are shifting, this sometimes involves severe injury or occasional death to losers in these struggles. I would certainly not intrude novel sounds in these situations without the guidance of experts on the behavior of wolves.

One of the many difficult decisions for people who provide care for large groups of captive wolves in limited areas is how much natural behavior to allow. Opinions vary from those who maintain that we should let nature take its course to maintain the vigor of the pack, to those who would avoid *any* potential injury to animals in their care. In some cases one might wish to try the careful experimental use of acoustic playback of calls to ameliorate problems that arise in captive circumstances for wolves. For example, recorded calls of other packs might be tried in order to turn captive animals' attention away from aggressing against each other.

## Artificial Prey

Carl Cheney, a pack leader and a friend since graduate school days, wrote a chapter for our 1978 book on *The Behavior of Captive Wild Animals.* I especially liked this chapter because it taught readers some important facts about predators, including their low levels of success when pursuing large and agile prey, and their critical role in maintaining the vigor of the prey species. Earlier in this book I described some of the dilemmas in trying to introduce live predation opportunities for captive animals. I will not belabor that point here, but simply remind you that where possible the best captive stimulation would involve opportunity for real hunting.

Unless birth control measures are employed, rates of reproduction in captive wolves are often high. In general, captive facilities that exhibit wolves as part of their attraction for paying customers and donors are seldom able to take additional wolves from sister institutions in which they become over-abundant. When there was a division in packs that resulted in increased reproduction, it became impossible to support all of the wolves in the Portland Zoo. Some of my students and I packed up a wonderful wolf in our van and delivered him to the wildlife station in Utah where Carl Cheney was doing some research with predators. The animals were regularly exercised in large arenas where Carl and his students searched for effective means of providing captive prey that would stimulate healthful activity in a variety of species of predators. Some of the ideas for the following recipes are based on observing the research of Carl and his students when wolves were the focus of their efforts. Other recipe segments are derived from successes that we have had in providing effective and apparently enriching activities for other kinds of carnivores. I believe that these methods should work well and safely with canids.

### Recipe 1. Simple, Inexpensive, Expandable

1. Make a "clothesline" out of nylon covered aircraft cable and motorize it with a reversible, variable speed, heavy duty motor. The clothesline should be long enough to extend over the wolves' enclosure at both ends.

2. Carefully select a clamp that is strong enough to hold parts of a carcass or other attractive food for wolves, and that will release the food when the predators tug on it.

3. Depending on the nature of the captive habitat, arrange the clothesline so that the attached food will move above the height of the tallest wolf but within easy leaping height.

4. Experiment until you find the best speed of movement to ensure a challenge to the wolves and prevent them from "capturing the prey" except on about ten percent of their pursuits. Do your best to

avoid cueing the wolves about when the prey is about to fly over. This might be accomplished by running the clothesline with the clamp empty at times.

5. Depending on available budgets, here are suggestions for adding variety to this enrichment:

A. If the enclosure is large enough, run more than one motorized clothesline across it.

B. Arrange the motorized end of the clothesline on a sturdy swivel base, and arrange several places around the perimeter to attach the pulley end of the clothesline to vary the angle of the path that the prey will take.

**Recipe 2. Moderately Expensive**

1. Select a number of practical locations at the perimeter of the enclosure where artificial prey might be animated for the resident wolves' entertainment, and where PVC tubes extending to feeding devices can be used to deliver food for successful prey "capture."

2. Decide on the easiest safe means to simulate opportunities to pursue prey. Here are a few suggestions that may help to keep costs down where budgets are limited:

A. Make one or more of the prey acoustic. Then limit the time of sound of the prey in a particular location and require that the wolves be detected at that location to receive food treats. This part of the enrichment can be easily and inexpensively done with just a single modified driveway type passive infrared motion detector and a time-delay relay.

B. Experiment with odors as indications of prey, devising a method by which the odors can be made finite in duration. If the wolves approach the novel odor within a pre-selected time, deliver food treats.

C. Get a sculptor to create some small artificial prey such as rabbits out of very tough resin-based material. Ask the sculptor to design these prey so that there are no crevices in which the wolves' paws will catch. Arrange a motor driven arm on which to mount the prey. Conceal all moving parts of the prey within an attractive naturalistic appearing enclosure with a dome type cover. Make a narrow opening that will allow the body of the artificial prey to emerge as it moves across the top of the dome. (figure 16-1). Carefully construct this installation so that you are certain that capturing the prey cannot injure the wolves.

Figure 16-1. Completed Artificial Prey Mounds Can Look Naturalistic

3. Use a series of feeder belts (see chapter 20) outside the perimeter, one for each prey location, with PVC tubing to deliver the food treats.

## Untried "Pregnant" Ideas

For wolves used in exhibits in zoos, I have always thought that it would be an excellent idea to include some sort of lengthy cave or other area that wolves might use as a den. In order to lend some excitement for the wolves and provide some memorable experiences for visitors, these darkened areas might be outfitted with a means for release of live prey or emergence of animated artificial prey.

This idea would involve the wolves disappearing from view for extended periods, so if the installation is made in a zoo or wildlife park, you would want to add a video camera and place a monitor so the public could view the hidden activity. You might want to place a second monitor in a viewing area for husbandry staff so that they could occasionally monitor the well-being of the wolves when out of direct sight.

**Recipe 1. Simple and Dependent on Local Acceptance and Regulations**

1. Adjacent to a perimeter area easily accessible from the outside, construct or dig out a sturdy extensive den where the wolves can retreat from view when they wish.

2. Occasionally release some small prey through a latchable perimeter trapdoor that opens adjacent to the open end of the den.

3. If the exhibit enclosure incorporates trees and/or dense vegetation this will help to make the hunt more lengthy and complex. (see fig 16-2)

Figure 16-2. Wolf Emerging to Chase Prey

**Recipe 2. Simple, Moderately Expensive, and Generally Acceptable to Visitors**

1. Same as Recipe 1, step 1.

2. Install a passive infrared detector (chapter 20) broadly focused on an area extending several feet from the den mouth out into the open area. Ideally this area should be at least 15 feet in length beginning at the den mouth.

3. Install a feeder belt and PVC tubing (chapter 20) just outside the perimeter fence and centered on the area being monitored by the infrared detector.

4. Install a protected outdoor speaker to play the sounds of some prey species. Use a solid state chip with the recorded sounds or other device of your choosing to produce the sounds. The speaker system can be installed in the same enclosure as the feeder belt if it is convenient to do so.

You can provide enticing varieties of food treats on the feeder belt and use an inexpensive computer with a simple program and I/O board to control the enrichment devices (figure 16-3).

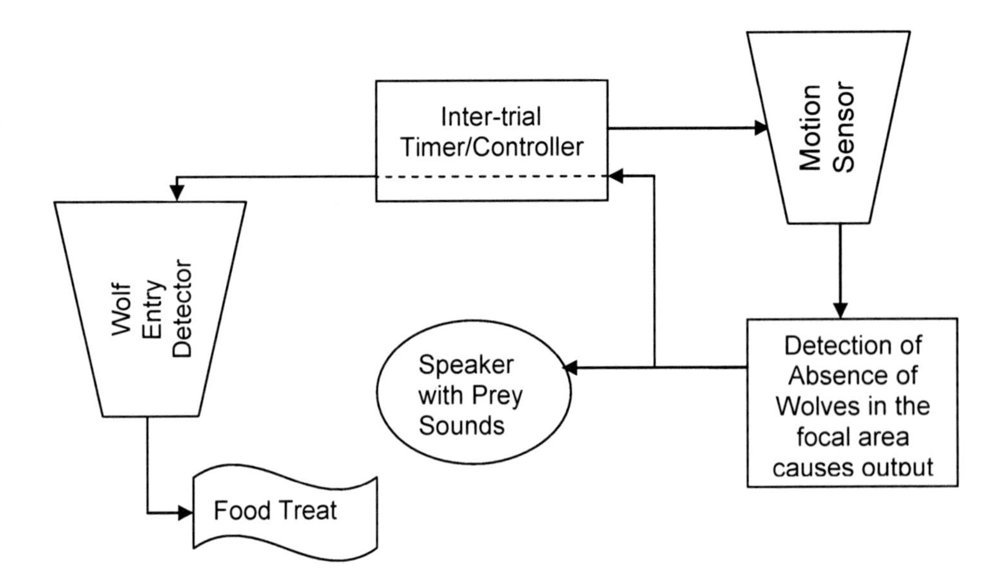


Figure 16-3. Simple Arrangement for Signaling Wolf that "Prey" are in Area, and Delivering Food for Prey Search

A. Program the time between hunting opportunities. It would be best if the time between opportunities was pseudorandom and within some minimum and maximum limits. For example, you might want to specify a minimum of a half hour and a maximum of an hour between hunts.

B. When this inter-opportunity time period concludes, switch on power to the relay contacts on the output of the passive infrared detector. Be certain that your program and device terminal connections are arranged so an output from the detector will only occur *when a wolf enters the area during a period that no wolf has been detected in the focal zone.*

C. When an output occurs from the detector, use it to begin the next step in a computer program. This step should be designed to do the following things: 1) turn on the sound of some prey species; 2) turn off the sound and operate the feeder belt to deliver a treat *when a wolf enters the focal zone;* 3) begin the next time interval between hunting opportunities.

If you have sufficient budget you can enhance this recipe by adding multiple potential feeding sites with associated detector and feeding mechanisms. Then, on each hunting opportunity, you could have your program randomly select which site would result in "prey capture."

**Recipe 3. Expensive, Generally Acceptable by Visitors, and More Naturalistic Exercise**

1. Same as Recipe 1 step 1.

2. Use a compressed air system to blow the odor of some prey carcass into the mouth of the den. This can be automated to occur on programmed demand by including a solenoid valve on the odor container.

3. Add a number of food conveyer belts and conduit at widely spaced and accessible perimeter locations away from the den. If feasible and affordable, vary the appearance, vegetation, hills, etc. in the enclosure at the sites of potential food delivery.

4. Place a narrowly focused passive infrared detector near each potential food delivery location.

5. In a protected area outside the enclosure, add a small computer or other solid state controller. Run conduit or remote controlled relays to deliver output signals from the computer interface to the solenoid controlling the release of odors and to each of the motor drivers for the feeder belts.

6. Program the computer or other control device to accomplish the following:

A. Control the time between hunting opportunities. It would be best if the time between opportunities was pseudorandom and within some minimum and maximum limits. For example you might want to specify a minimum of a half hour and a maximum of an hour between hunts.

B. When a hunting opportunity is initiated by the program, generate a signal to release a whiff of an attractive odor near the den mouth.

C. About twenty seconds after the release of the odor, have the program scan the state of the passive infrared detectors to see which, if any, have wolves nearby.

D. Have the program randomly select among the potential feeding sites that are not currently occupied by a wolf.

E. When a wolf is detected entering this randomly selected site, deliver food in that location (fig 16-4).

For some institutions, where visitor opportunities to see wolves are a major priority, it is an easy manner to add a motion detector in visitor areas, and have the presence of visitors increase the probability of a hunting opportunity occurring while they are observing the wolves.

Figure 16-4. Wolf Searching for Prey

**A Parting Thought About Wolves, Coyotes, and Other Predators Destined For Reintroduction**

It has become abundantly clear to those involved in reintroduction of endangered species to the wild that it is necessary to do some pre-training for animals that are bred and initially reared in captivity.

Otherwise they may never learn that it is necessary to escape from natural dangers or how to effectively compete for food in order to survive when reintroduced to the wild.

In the case of predators close to the top of the food chain, there are additional concerns that may require some opportunity for the predator to learn skills necessary for survival. This is especially true when they are to be introduced to areas where humans have settled. Captive animals are most often habituated to humans who treat them kindly, or from whom they are protected by artificial barriers. Encounters with humans upon release will not always be safe for these animals.

If you intend to successfully reintroduce wolves or other mammalian predators to natural environments, I believe it is important that while in your care they learn to hunt natural prey and also learn to avoid taking domestic prey. A creative program might also include teaching these predators how to avoid poisonous plants and other dangers.

The problem that exists for those who might attempt this approach is that they will inevitably encounter opposition from extremists on both ends of the issue about what is appropriate. On the one hand, I have met many ranchers who will tell you that the "only good predator is a dead predator" and will resist efforts to support effective reintroduction programs. Simultaneously, there are many devoted animal lovers who are of the opinion that there should be no support for any procedure that intentionally inflicts any discomfort on captive animals.

I have a few suggestions for those brave enough to give it a try even in the face of formidable, and often wealthy, opposition. You might want to be certain to combine some opportunity for predators to encounter and, ideally, to sometimes capture *living* prey. Remember that predators have a low success rate in nature, so do your best to make capture a significant challenge in captivity. Provide hiding places for the prey species, and have the predators learn that they have to really work to feed themselves. If you need to conserve every animal for reintroduction, there will be a lot of judgment calls for you to make. You cannot directly mix animals that must survive with potentially lethal enemies that they might encounter in nature. You cannot allow some of the predatory animals to socially acquire knowledge about which plants are poisonous by watching their conspecifics become ill or die from eating the plants. Some approaches to try on a careful experimental basis: 1) Place the potential natural enemies just out of reach of the animals in your care and vice versa. By this means young wolves, for example, might get the message that these enemies would like to harm or eat them. 2) Use carefully measured or treated samples of poisonous plants likely to be encountered in nature. By this means, guarantee that there is not enough toxic material to cause death, but enough to cause mild illness that will teach the animals to avoid it in the future. 3) Consult the literature for works by folks like John Garcia and Carl Gustavson to examine the feasibility of using conditioned aversion techniques to help prevent predation on domestic animals. These techniques combine the smell and taste of the flesh of the species to be avoided with agents such as strychnine to produce transient illness in the potential predators. If properly accomplished, this method can be of significant value in keeping wolves from being destroyed because they are seen as rogues which attack domestic animals. While some readers may find this approach unpalatable (pun intended), others might see value in using extreme methods such as this to deal with problems that may have no other workable solution. An advantage of using flavor aversion techniques rather than other negative reinforcement methods is that taste aversion produces longer lasting effects and greater resistance to extinction (Gustavson et al 1978).

When working with captive animals destined for reintroduction, enrichment efforts should focus on preparing the animals to have a real chance to survive inevitable challenges when released.

# 17

# RATS AND MICE

While my original plan was to write a chapter entitled "Rodents," I soon dissuaded myself from this idea. That is indeed a better title for a series of books than a chapter, since more than half of the mammalian species alive today are in the order *Rodentia*. These mammals range in size from the pygmy mouse which weighs only 4 grams to the capybara weighing more than 50 kilograms. There is a description of some work we conducted with capybara in my earlier enrichment book (Markowitz, 1982). Here I will limit myself to rats and mice, with most attention paid to laboratory animals that are much in need of enriched environments. There is evidence that the animals will benefit, *and* that biomedical research efforts that employ them as models can be interpreted more meaningfully if opportunities for richer lives are provided to the captive animals involved (Markowitz and Timmel, 2005).

I have chosen to deal exclusively with enrichment ideas for rats and mice in this chapter for two principal reasons. First, these are the species that have the greatest number of individuals in captivity, and about which we have the most applicable data. Second, virtually nowhere is there more apparent confounding of our judgments as a function of historical accounts than with these animals. We humans owe a great debt to rats and mice for having served so well for so long in our research efforts. There is an important distinction between rats and mice that are bred to serve humans, whether in research or as companions, and those that spread disease in impoverished human domains. Paradoxically, when the first contemporary regulations requiring more humane treatment of mammals in research were issued, rats and mice were specifically excluded from the definition of "animals." One can understand the practical concern that costs involved in effective enrichment procedures for the enormous population of rodents used in research could be prohibitively high for many institutions, but there is certainly no reason to exclude rats and mice from protection laws on the basis that they are unfeeling or totally incompetent to learn anything..

Returning to our earlier consideration of the definitions of enrichment, there are abundant studies with rats and mice that show that enrichment can be effective in causing positive changes in their captive lives. Most of the recipes in this chapter are based upon procedures which have been observed to result in rats and/or mice using the new opportunities provided them. Consequently, the criterion of use by the target species is clearly met. Comparison tests provide evidence that specific kinds of enrichment tend to be preferred over other choices. Therefore, inferences about the animals enjoying their new opportunities are as plausible here as they are for species for which enrichment is mandated. If we use criteria associated with the production of more naturalistic behavior, many of the ideas put forth in the recipes in this chapter are designed to provide greater opportunities for engaging in species-typical behavior of the kind apparent in wild rats and mice.

Rats and mice have not often been glorified except in animated movies, cartoons, and children's stories. These are clearly fables meant to be especially attractive to children, and part of their popularity rests on commonly held notions that it is ludicrous to expect that rats or mice might in any sense have the mental abilities to engage in complex behavior. Before embarking on the recipe sections of the chapter, I will close this chapter introduction with observations about findings that belie the notion that rats and mice are "without feelings," without intelligence and generally insensitive to changes in their lives.

Current technology does not allow for definitive proof of sentience in any non-human species. Yet we are inclined to judge by inferences from observation and research results that animals of many "advanced" species have feelings. One frequent basis for judgments about mental processing of information in non-human species members is ability to learn solutions to problems that seem *to us* to require reasoning. Anyone who has read introductory books on animal learning knows that rats and mice can solve surprisingly complex mazes and improve their performance based on previous experience.

We sometimes judge whether species are sentient based on observations of what *we* consider individuality of behavior (sometimes labeled "personalities") in species members. Not only are "intellectual" differences apparent in rats and mice, but the heritability of behavioral differences and abilities has also been long established (e.g., Tolman 1924; Tryon 1940; Markowitz & Sorrells 1969; Markowitz & Becker 1970).

Another anthropomorphic criterion that is employed concerns the presence of what *we* judge to be sensitive expressions of love in other species. For example we tend to judge how much they "love" their offspring based upon apparent pain or sacrifice parents are willing to endure to ensure the well-being of their progeny. In formal tests, it has been shown that mother rats will undergo much more pain to retrieve their infant offspring than they will endure to get food when they are extremely hungry or water when excessively deprived of drink (Warden, 1931). Although some will dismiss these qualities as simply attributable to genetically maintained automatic specialization that serves "selfish genes," this is much too glib a suggestion for my taste. No matter what the heritability of specific behavioral attributes, our usual anthropomorphic judgments about the feelings and sensitivities of other animals are based on assessment of their behavior. Rodents show wariness. Rodents show considerable parental care. Rodents make sounds of suffering when they are in pain. They deserve our respect and attention in providing as comfortable lives as we can afford for them when they are serving human purposes.

## Playgrounds for rats and mice

A half century ago I worked with a team of scientists at the University of California, Berkeley who studied relationships between brain chemistry/anatomy and behavior. Among research findings was the remarkable role that enrichment could play in changing the behavior, brain chemistry and anatomy of rats (Bennett et al 1964; Tapp & Markowitz, 1963). In one part of this research, rats were separated into carefully matched groups. Rats in one group lived rather isolated lives in laboratory cages. Members of the other group were released into large "playgrounds" each day for significant periods of exercise and interactive play. Based on results of this research, I am certain that the first simple recipe will work in enhancing the lives of rats. I have confidence that providing this sort of enrichment will also prove stimulating for mice.

**Recipe 1. Simple, Cheap, and Fun**

Get some small pieces of hard wood and entertain yourself (preferably while listening to Mozart) by inventing and building playground equipment for rats and/or mice. Among the simplest kinds of equipment that they will certainly use are small climbing structures, tunnels, and wheels that they can rotate with their paws.

If you are working in an institutional setting, be prepared to fight the good fight with the veterinary staff and inspectors. They will almost certainly prefer that you use the commercially available plastic mouse houses and manipulable toys that can be washed in a cage washer. Rats and mice will appreciate your efforts to get them softer, more natural materials that they can gnaw on and that are not totally unnatural surfaces. Although you may have to replace them occasionally, good construction techniques and proper selection of wood will make it possible to clean your devices occasionally in antibacterial soapy water and to use them for a long time.

If your facility has a limited enrichment budget, and the population of rodents is known to be free of serious transmissible disease, using the same playground for groups of conspecifics at different times of day may provide the additional stimulation that comes with smelling animals other than one's cage mates. If the rats or mice in your care are used in any work involving breeding, it will be necessary to exercise extreme caution to avoid introducing pregnant females to the smell of unfamiliar males. It has long been established that the smell of alien males is sufficient to cause loss of pregnancy in females (Bruce, 1963).

**Recipe 2. Simple, Inexpensive, and Low Maintenance**

Do the necessary literature research to identify the safest and most reliable kinds of running wheels for the species in your care. Among the important considerations are that the wheel be sanitizable and that the device should not include mesh in which the rodent can catch its paws while exercising.

Since the demands on space in many research facilities are high, it will be necessary to convince those in charge of budgets and space assignment that providing cages with running wheels is a necessary component of humane rodent care. Study of the literature will provide information to bolster your position in seeking larger budgets for enrichment, including studies showing how species-appropriate exercise may enhance the meaningfulness of collected data (Markowitz & Timmel, 2005). There are many manufacturers who can include relatively low cost useful additions to running wheels such as revolution counters to automatically record extent of exercise. In more expensive and ideal systems, electronic counts of revolutions and time of exercise are automatically transmitted to a computer for analysis.

Some facilities have lots of animal care providers anxious to help with enrichment efforts. One way to lower equipment costs is to outfit as many large cages with running wheels as you can afford, and to rotate animals from their home cages into this environment for an hour a day.

**Recipe 3. Very Inexpensive, Especially Useful for Mice**

There are a wide variety of manufactured plastic "mouse-houses" that can be purchased even with small enrichment budgets. I personally prefer building my own out of less artificial materials, but in any case there is very little doubt that mice will happily retreat to these places on a frequent basis. If the criterion for enrichment in your facility is frequency of use of newly provided items, this is a virtually guaranteed method of success.

However, it is important to remember that there may be significant effects of routine housing procedures on research outcomes (Markowitz & Timmel 2005). One especially pertinent example with respect to mouse houses arose in some of our enrichment efforts at the University of California, San Francisco (UCSF). A colleague dedicated to providing better lives for her research animals asked me if there was anything that I knew that mice would absolutely like and use in their home cages that she could afford with a limited budget. With little hesitation I replied that mice liked the opportunity to cuddle in mouse houses. Later she told our enrichment crew in some despair that although the mice did indeed readily use the mouse houses that she purchased, she had to remove the houses from their cages. Her research staff discovered that mice provided with mouse houses exercised less than mice in barren cages. Consequently their recovery from spinal damage (the major focus of the research) was significantly retarded.

Findings such as the one described above present difficult conundrums for enrichment workers *and* researchers. Some readers will be asking themselves whether this means that the mouse houses reduced activity below levels that were healthy for mice. An equally relevant question is whether exercise of unusually low levels compared with wild mice occurs for animals without anything else to do except wander around their small cages. Deciding whether something is "enriching" for animals is based on rather arbitrary criteria. The methods employed to reach enrichment goals may have serious impact on other aspects of animals' lives and clearly affect many research outcomes.

In biomedical research other mammals are frequently used as research subjects, and the results of these efforts are used to assess the probable effectiveness of the same procedures in treating humans. If the laboratory animals employed have lives with none of the environmental qualities which free-ranging members of their species typically encounter, and are largely powerless, can they be considered behaviorally healthy? Even when the research focuses on physiological and anatomical *outcomes* of various experimental procedures, there is good reason to use behaviorally healthy animals as "models". Environmental qualities and husbandry procedures can alter an animal's physiological and anatomical state (e.g., Bennett et al 1964, Bruce 1964, Tapp & Markowitz 1963).

When biomedical researchers are trying to meaningfully predict the probable effect on humans of a medical procedure or drug regimen, it behooves them to collaborate with husbandry personnel in maintaining *both* the physical and behavioral well-being of their test subjects. Using powerless animals with no opportunity to engage in normal behavior for their species will significantly reduce the chances that results will be the same when the methods are applied to humans who typically live richer, more behaviorally healthy lives. Socially housed animals that are given some control of things in their environment are better "models" for predicting likely human results of a procedure.

**Other Brief Ideas for Developing Your Own Enrichment Recipes**

1. Joinable plastic tunnels and other kinds of "trails" appropriate for mice or rats can be purchased or rather easily made. They can be used to allow animals to move between cages if sufficient room and caging is available. A caveat is that some strains of rats and mice are notoriously aggressive while others are more gently social. If rodents are to be provided opportunity to "visit" each other, you need to ensure that they are compatible and will not injure or kill one another.

2. To generate more activity, consider providing them a means to order food in their home cage when they wish. This can also serve to accustom laboratory bred mice or rats to doing some work to earn their food as may be required in experimental efforts. Although it may only be possible where maintenance budgets are above average, a simple procedure would be to provide living quarters with solid or liquid food dispensers and levers or buttons for the rodents to manipulate to deliver measured amounts of food or water. There are relatively inexpensive solid state controls that can be used to generate response requirements such as pressing a lever several times to produce food.

3. You will find that most rodents will use exercise wheels routinely without coaxing. You can give them a little more control over their own lives by using the exercise wheel to activate a food delivery mechanism that is attached to their cages. Most running wheel suppliers have options for electronic tabulation of number of wheel revolutions for output to computers and/or counters. This output can also be used to operate a food delivery device that allows you to select the response requirement (number of wheel revolutions) for delivery of food.

A less expensive option for those who have more time than money is to attach some protuberance to the running wheel that will cause a response to be registered each time the wheel turns one revolution. For just a few dollars, you can do this by making the protuberance the permanent magnet that operates a reed relay. These units are readily available since they are commonly used in monitoring the opening and closing of doors or windows. The relay contacts can be used to operate your feeding system. Since the permanent magnet component is water-resistant and requires no wires running to it, the wheel can still be removed for repair or cleaning with no trouble. The reed relay component comes sealed but with bare wires protruding on each end to allow connection to the sealed relay contacts. Don't forget to use shrink tubing or epoxy to waterproof these wires after you make your connections.

4. Consider combining enrichment for animals in their home cages as components of your research designs. In this manner you may provide ways for animals to live in richer environments and simultaneously enhance research. For example, in many discrimination studies that use non-humans as subjects, it is feasible to provide automated discrimination apparatus in the home caging allowing animals to learn the contingencies any time of day. In addition to providing regular stimulation for the animals, this will reduce demands on staff time to run scheduled pre-training sessions. Some investigators have conducted entire research projects by attaching apparatus to the home cage and electronically tabulating the animals' success and failure in making discriminations that led to automatic delivery of reinforcement.

It is often possible to provide much more adequate budgets for enrichment by incorporating enrichment equipment as a necessary component of your research procedures. This is because the major research funding agencies are much more likely to provide money for enrichment apparatus when it is made clear that the apparatus is an integral part of research efforts.

In anticipation of the next chapter, it is a good time to mention that rats (including some of those that are retired from serving as research subjects) can be especially nice pets. They can express affection by running up and down your arms and snuggling in the crook of your neck. This is especially endearing for those of us who are ticklish. Rats are easily fed, neatly present their feces in nice solid primarily dry form for clean up, and cause only minimum inconvenience for friends who tend them when you are away. And, as has been suggested in this chapter, you can have fun making and acquiring playground toys and other enriching items for them without breaking your budget. Some research facilities have developed "adopt a research veteran" programs to place cats and dogs with people who will enjoy their company and provide humane care for ones they adopt. Non-human primates that are used in research are often sent to rehabilitation facilities or retirement sanctuaries where they can live richer lives. Rats and mice that are used in research far outnumber the other species employed in such efforts. They are also highly reproductive and have a gestation period measured in weeks. Consequently, it would require a prohibitive amount of work to develop parallel programs for them. However, for those rodents who are tame, non-aggressive individuals, it is rewarding to give these "research veterans" a loving home for the rest of their relatively short lives. I can attest to that from personal experience.

# 18

# PETS

Do you ever bite your lip in order to prevent laughing directly at someone??? I vividly remember one such occasion when a very famous and wealthy woman came up to the podium following a public lecture to thank me. She told me how much she agreed with me that captive animals were in need of humane treatment. While leaning closer to make her point in the noisy outdoor auditorium, she pulled on her dog's choke collar to keep him in line. He was an especially attractive dog with ears clipped to be fashionable for the species, and showed responsiveness to her commands for obedience.

How can it be that some of those who vigorously oppose the use of captive animals in educational and research efforts simultaneously enslave their companion animals? If I have offended some readers with this question, I apologize....but at least I have your attention and your adrenalin pumping. Throughout this book I have tried consistently to convey the notion that the best kinds of enrichment are those in which we empower others and relinquish some of our control over their lives. Nowhere is the need for this more evident than can be seen by observing our behavior with respect to those animals that we love most dearly. We "want to be proud of them," "want what is best for them," "wonder why they do not know (as we implicitly do) what is best for them and how to behave properly."

There is certainly a need for us to spend more time carefully empirically evaluating what will make animals most happy and able to control more aspects of their own existences. Whether we are talking about pets that provide us comfort as companions, that serve as helpers with our disabilities, or that bring pride because of their winning ways at pet competitions, if we truly love them, humanity cries out for us to provide pets more power to choose the activities that are attractive for them and not to simply make them behave in ways that we select for them when it is possible to allow them to do so safely.

This chapter will concentrate primarily on recipes most pertinent for dogs and cats. However, I hope that some readers with other pets may find utility in recipes in this and other book chapters. Perhaps they may help you in designing of enrichment equipment or procedures that empower your companion animal. Study of published research of species-typical, or strain (breed) dependent behaviors of free-ranging animals of the same type as your friend will be of help in your efforts. Why not study what the pet you love might wish to do if given the chance to choose for themselves? You may find that other pet lovers may find the techniques you devise of benefit if they learn about your work.

## Ways to Entertain Your Pet when You Are Not at Home

### Recipe 1

Some of you may occasionally leave the TV or Radio on to entertain your pet. How do you know whether they like it on? How do you know if they get tired of it, even though it may at first attract their attention? Why not let them decide whether they want it on, and for how long?

Arrange any of the following ways for your pet to turn on the TV or a radio. If you want to be really gallant, provide them a means to select which device they wish to use for entertainment. Here are some simple suggestions for devices you might want to use in providing your friend the chance to choose:

1. Install an inexpensive passive infrared motion detector (chapter 20) focused on a comfortable area that they can use for the entertainment. By using the control dial for adjusting output duration that is part of most motion detectors, you can easily make adjustments for how long the TV or radio will stay on when the animal leaves the area.

2. Install a switch that you can teach them to operate with their paws to turn the device off and on. You can either use a push button switch that toggles back and forth between "on" and "off" each time that it is pressed, or you can make a panel with separate switches.

3. If you have a remote control for the device, you can make a sturdy little case that gives them access only to the controls you wish to allow them to utilize, and teach them how to use it (see Recipe 2 for additional related ideas).

4. Get a voice operated relay, and have them bark or purr to turn the device on or off. Consider however, that if your dog likes the fun of barking to control the device this may not endear you to proximal neighbors!

### Recipe 2

It will take some discretion on your part to decide what functions of the remote control you can logically allow your animal to access when you are not home. (This will be immediately clear to those of you who have children and have had neighbors complain about loud rock music emanating from your house when you are away.) Select one of the following two options based on factors such as the size of your animals and their dexterity. For example, you might not want to give a Saint Bernard direct access to the delicate buttons on your remote control because the sheer size of their paws might preclude them from discretely selecting particular functions. On the other hand if you have a clever Chihuahua and you follow option 1, it may work just fine.

1. If you have a small dog or cat with sufficient dexterity to operate the buttons on your remote control, make a suitable container that can be fixed in place where you wish your pet to have access to the control. This container can be constructed rather easily by making a template that has holes in the locations of the buttons that you wish to allow the pet to operate and then drilling holes large enough for access. If the remote control has very tight spacing, you may want to make "control caps" that protrude to the outside and that have stems in contact with the buttons on the control.

My recommendation is to allow your pet to control channels pressing the arrow keys that enable this function, and that you provide access to the power button so that they can turn the TV off and on. Of course, I know that some of you are certain that your pets are brilliant beyond belief. In such cases, you can always offer your gifted companions a smorgasbord of opportunities to use various audio and video devices, or to play computer games.

If purchasing a radio for your pet to control, select one with a remote control that allows both station selection and silence as choices. Some controls are programmable and may allow you to preset a number of buttons so that your pet can easily select the genre which best suits their taste, and still precludes your having to listen to talk show drivel.

2. For larger animals, there are no really straightforward options, but I will make a few suggestions for your consideration. The first is adventurous. It involves taking apart your remote control to see how it works, and finding a means to close the contacts with much sturdier and larger buttons appropriate for your pet. Unfortunately, today most cheap mass manufactured controls are not easy to modify in this manner because they use a single special pad as the contact plate for all of the individual buttons. It requires real care, and some expertise, to rewire such controls.

An alternative is to get an expert in electronics to show you how to control the required functions with remote relays, or specially coded outputs that operate the infrared remote controls of the TV or radio.

If you are a good craftsperson, or can woo one to help you, another possibility, is to make very sturdy plungers that exert only the minimum required excursion force on the keys to reliably operate remote control functions. This can be done by using a spring-loaded arrangement to allow your large pet to depress the plunger until it reaches a stop that guarantees that she/he cannot break the remote control when selecting between options.

While you may think such recipes require considerable time and effort on your part, such work on behalf of your companion will involve much less time and energy than your pet spends in efforts to please you. You will be rewarded in many cases by the surprise that your clever companion shows at their new found abilities once they identify the contingency between their behavior and changes in the TV or radio output.

**Recipe 3**

Make a recording of yourself saying pleasant familiar things to your pet. Arrange a motion detector or other device of your choice so that your pet's movement into some particular area will turn on the sound of your voice for as long as he or she remains in that area. Alternatively, you can simply use a plug-in timer to occasionally turn on your dulcet tones. In any case, you will want to monitor your pet's response to this recipe. If you are gone for long periods you would not want this recipe to backfire and have your friend wail in longing for the real you. Additionally, if you try this simple recipe, you may want to stay home and watch from the next room when you first employ it, both for fun and to make sure that your pet does not go nuts searching for you.

**Recipe 4**

This is identical to recipe 3, except this time use a DVD player to allow your companion the exceptional pleasure of seeing you as well as hearing your voice when you are absent. (Be sure not to beg them for a trick or other behavior in your recording, unless you also rig up some way to identify their appropriate response and reward them properly.)

## Other Brief Easy Recipes for Entertaining Your Pets

### Recipe 1

Just like young children, some pets are excessively afraid of humans other than their "masters." One way to have them become more approachable and friendly may be to show them video recordings of other human beings acting kindly toward animals. Since there remains considerable controversy about the extent to which animals commonly kept as pets perceive videos in the same manner we do, this will require some observation on your part to evaluate the effectiveness.

A first step would be to make or find some video recordings of humans being kind to pets and watch them with your companion. If they show interest, then try the simple step of providing them a means to turn on this recording when they wish.

Another use of the same recipe would be to show a video of animals being friendly to each other in order to reduce the apprehensive nature of pets that have never spent time with anyone but you and their mothers. One of the places that I frequently hike is an area where pet lovers, including people hired to walk dogs, bring dogs that obviously have a great time safely meeting and interacting with other dogs.

### Recipe 2

This recipe is especially good for cats. Puzzle boxes have been used for nearly a century to evaluate the learning abilities of domesticated cats.

Have fun inventing complex latches for puzzle boxes in which you place treats for cats. Let them entertain themselves by fussing with the latch mechanism until they learn to open it and obtain the prize. Make a big variety of these and randomly provide them periodically to cats. See whether the problems always "seem new" to them, or whether once they have learned to open a particular mechanism they remember it the next time they choose to open the box. This can be more fun and less entangling than a new ball of twine!

### Recipe 3

Try inviting your pet to get on your motorized tread mill with you, or get them one of their own and place it next to yours. This idea will be especially useful if you and your pet tend to get insufficient exercise and/or to overeat.

### Recipe 4

Use your ingenuity and build something that requires your over-nourished and under-exercised pet to get some exercise in order to obtain ***small*** food treats or other rewards. For example you could build a modified "rooting device" of the sort described in earlier chapters, so that your pet could do some work to crank out the reward. (Of course you and your pet might better benefit from taking a daily run together.....but you know that you and your pet are too slothful to always be counted on for that....and taking some pets on a run is problematical.)

**Recipe 5**

We have found this procedure especially useful in providing enrichment for kittens and puppies and encouraging healthful exercise. Go to a good toy store and get some giant non-toxic material building blocks. Make a three dimensional labyrinth that your pet can climb in and out of and jump between levels to explore. Change the configuration frequently to introduce novelty to keep it more entertaining for your pets to explore the labyrinth.

**Recipe 6**

If your pet would enjoy living part of its life in your backyard or other outdoor area, but you are concerned that they might run away, arrange a controlled run for them. Install a ring and line arrangement and provide an extensive exercise area. This will allow significant exercise while keeping them safe.

Be certain to carefully construct and test your arrangement to make certain that there is no chance that your pet will wind up suspended in air because of entanglements in the line.

**Recipe 7**

Out of choice or necessity, some of you may always keep your pet indoors. In one such case, I was astonished to watch a cat merrily using the extensive enrichment that her owner had invented for her. It was a series of cat-sized steps running up the walls of the living room. This provided elevated paths which the cat could ascend to look down upon inferior humans. This pet used her perch quite imperiously

.

I hope and expect that all of you will be highly critical of the simple-minded nature of this chapter and the paucity of truly innovative techniques. This will mean, of course, that you are either already way ahead of me in providing means for your pet to control parts of their own daily opportunities, or that you can think of a thousand better ideas. Either would please me greatly, and I look forward to your scorn at my skimpy ideas.

# 19

# Potpourri of Ideas for a few of the Kinds of Animals not Covered Elsewhere in this Book

This is not a detailed recipe chapter as are most of the others in this book. Rather it is just a collection of a few notions based on some experience with particular kinds of animals for which few enrichment efforts are typically made.

## Farm Animals

There are seemingly endless opportunities to improve the lives of farm animals. Improvements are increasingly coming into play, but implementation depends on the budget available and the kindness of those involved, including consumers. Witness the number of "free-ranging" animal food products appearing on our shelves, and the increasing efforts to lobby for humane treatment of farm animals. While farm animals by definition lead restricted lives, we can surely take measures to make their daily existence more pleasant. Here are a few ideas developed in zoo farm exhibit settings that seemed meritorious to me, along with a couple I would propose.

Historically, a great number of zoos have done their best to acquaint urban kids and their families with the interesting and complex nature of many farm animals. In some of the neatest inexpensive exhibits, there were peep holes to allow the public to unobtrusively peek into the living quarters of critters ranging from chickens to horses (*Equus ferus*) . In the best of these captive homes/exhibits, there were always members of the staff present to ensure that respect was given to the animals and that people did not intrude on their activities. While poorly supervised "petting zoos" may help to endear domesticated animals to people, these exhibit areas often do not teach anything about the lives of animals on farms or ranches. In the worst cases, they may teach kids to disrespect animals or to fear butting rams. I have sadly observed incidents where demeaning ideas about particular species were reinforced because of inappropriate comments by staff or volunteer "mentors." In many children's zoos around the world, I have observed unattended visitors harass animals and/or make fun of the lot of the animals with which they were interacting.

In contrast, there is growing effort by more enlightened zoo staff members to enrich the lives of animals in farm animal zoo exhibits and to provide the animals enriched opportunities to engage in species appropriate behavior. A few websites include ideas for enrichment for farm animals. A great many of the suggestions posted on these websites emphasize the importance of simply introducing wide varieties of browse for animals. It is my experience that many markets are willing to donate fresh foods to provide feeding variety for animals in zoos and sanctuaries. For some creatures, such as potbelly pigs which are increasingly displayed in zoo areas primarily designed for children to visit, it has been found that burying produce and allowing the animals to root for it may be stimulating and help to encourage more naturalistic behavior.

Useful inexpensive additions to quarters for cattle include mounting broom bristles on the walls of enclosures to allow the animals to scratch their backs, hanging large balls of tough material that they can bat around with their noses, and presenting sturdy plastic garbage cans that they can kick around to entertain themselves. (This last method of providing entertainment opportunities has proven somewhat more exciting for horses than for cattle.)

I am one of those eccentrics who sometimes stop at a farm fence to visit with cows while eating my lunch. During some of these visits, I dreamed about ways in which we might utilize gadgets to improve their opportunities. But when I recovered from these daydreams, I always concluded that providing large areas for them to freely roam and graze was the kindest thing that we could reasonably do. I continue to believe that encouraging ranchers and farmers to move toward free range husbandry is the best strategy for enriching the lives of cattle raised for food. Of course, as we humans increasingly overpopulate the planet and consume more and more, this task becomes formidable and may eventually become impossible. The demise of small family farms and the increase of "factory farms" provide testimony that the days when most farmers and ranchers knew, respected, and individually cared for the animals who provided them milk and meat are largely in the past.

Not surprisingly, horses have received more individual attention than other farm animals. This is to be expected because of the number of people who have learned to love them as companions and more than just creatures to ride while performing duties on farms and ranches. Among the simple things that have been found stimulating for horses are reducing the amount in each feeding while increasing and varying the number of feedings, and getting them out to pasture as often as possible. Toys and treats of varying kinds are available for horses. These range from scented plastic apples to knotted puzzles for them to untie.

Because so many horses are essentially pets, it is not surprising that their owners are certain that the tricks they have trained them to perform on command are greatly entertaining to the horses. I confess that my bias runs in a different direction, which you might infer if you have read earlier parts of this book. I would rather see horses entertain themselves when they wished rather than when it fancied a human to get them to perform. Many of the recipes that I have provided for animals such as camels and zoo hoof stock (e.g., allowing them to feed themselves in interesting and mentally stimulating ways) could be used in planning enrichment for horses.

For goats (*Capra aegagrus*) in limited environments, keepers at a number of zoos have found that inverted buckets provide places for them to climb and that goats regularly use them for this purpose. It should be possible, even on limited budgets, for flatland goat farmers to seek some way to incorporate some places for elevation within the areas where these animals are maintained.

Chickens (*Gallus gallus*)may benefit from opportunities for sand bathing and will also greatly enjoy wider varieties of food than they are ordinarily provided. Some of these foods can be grown or bred at very low cost (e.g., crickets, mealworms, and hardy vegetables such as kale).

I have devoted a previous chapter to enrichment for pigs, common farm animals that are so capable in learning new things that it seems certain to me that they suffer in unstimulating environments. The paucity of enrichment efforts for pigs is doubtless in part a product of our customs in disparaging people by calling them "filthy pigs," "lazy as swine," etc. How come we do not typically refer to the person who is clever enough to take the time to enjoy a mud bath as "clever as a pig"?

Encouraging young farm and ranch kids to invent methods to enrich the lives of animals in their care, rather than to haze animals, might produce exciting results. Youngsters love to feel that they are participants in work of their own invention, and they often come up with more inventive ideas than their older kin.

## Insects and Spiders

While there are a few really neat insect zoos in various places, it seems to me that too many institutions with such displays go overboard in graphics, preaching to people about the importance of these animals in ecosystems, and trying to allay phobias and fears. It would be wonderful to see increasing emphasis on providing opportunities for direct observation of the normal living and working conditions of insects and spiders. I do not subscribe to the notion that animals lacking highly evolved cerebral processing equipment *think* about their state of affairs. But it seems to me they would live a more comfortable existence performing the behaviors for which their genetic endowments have provided the equipment. So, I would love to see more live opportunities to witness what so many wonderful nature films have shown about the lives of insects and spiders.

Some ideas already widely employed include: providing hives and/or nests with transparent sections that allow visitors to observe the social behaviors of various insects, using natural foraging materials for various kinds of ants including carpenter ants, and providing means for observing behavior of spiders and insects without disturbing the animals at work.

Forty years ago, I designed an exhibit to allow insects and spiders comfort while engaging in natural behavior and provide zoo visitors opportunity to easily watch these animals in action without disturbing them. This plan, designed to effectively integrate this with the teaching mission of zoos and museums, specified optical magnification systems and amplified audio to provide a live "inside look" at the world of some small species on a large screen. This can be especially exciting for those trying to anticipate what the "next move" of the insect or arachnid will be. My design was not implemented because of a zoo budget shortfall. However, today there is increasing use of techniques like this in zoos and museums that exhibit live insects.

## Penguins

Late one afternoon I stood in an observation shack in New Zealand, gazing through a narrow slot, waiting for the first group of Yellow Eyed penguins (or "Hoiho" *Megadyptes antipodes*) to return from the sea. Spectacular behavior can be observed when these birds are swimming and leaping to shore. My patience was rewarded when I got a clear view of the first penguin to return. He awkwardly scrambled onto the shoreline rocks at the foot of a beachside hill. His skill in avoiding being dashed on the rocks was more impressive than his awkwardness was disconcerting, while I observed his behavior, and empathized with his efforts. I wished that others could have been with me to see the beauty of this event, and thought how misleading the typical captive penguin habitat is in displaying their behavioral capabilities.

The penguin waddled his way up the hillside, finding a resting place in the form of a ledge several meters above the surf line. Here he impatiently scanned the sea watching for the return of other penguins. An hour or so later they began to let themselves wash onto the rocks, and the penguin "scout" began to waddle down the steep hillside to join them. Observing this whole series of early evening events, I could not help but anthropomorphize about what was going on in this beautiful bird's brain. Those who see nonhuman animals as pre-programmed creatures would have to do a great job of circumlocution to avoid the conclusion that this penguin was happy to see his mates and join them.....in spite of the fact that his brain was neither convoluted, nor as cognition-capable as our brains.

There are 17 or 18 living species of penguins, all in the Southern Hemisphere and ranging from tropical to Antarctic in their primary locales. Besides their varying habitats, the size of penguins varies over an exceedingly wide range, from Emperor Penguins which grow to about 1.1m (3.7 feet) and weigh as much as 41kg (90 lbs), to Fairy Penguins (also called little blue penguins) (*Eudyptula minor*) at 41cm (16 inches) and about 1kg (2.2 lbs). While all are built for swimming with wings modified into paddle-like flippers, life styles and species-typical behavior are quite diverse. For example, some mainly feed on fish and some on krill; Antarctic penguins show the unique behavior of "tobogganing" over ice on their bellies while propelling themselves with flippers and feet; and other species show special behavioral adaptations to best survive in their habitats.

Formal studies of penguins have revealed that their hearing is good and that their eyes are adapted for underwater vision, leaving them near-sighted when emerged. They have color vision and a limited sense of smell. There is evidence of vocal communication and mate recognition.

Watching penguins swim provides further evidence of diverse capabilities. Since they are air breathing and can be quite fast underwater, it is not surprising that some species have developed the adaptation of porpoising in and out of water with graceful leaps. Yet a visit to most facilities that house penguins will yield little evidence that they are encouraged to exercise their amazing adaptations. This does not well serve the mission of using these animals for conservation education, nor does it appear that penguins are stimulated sufficiently to provide them very pleasant lives. Here are a few ideas to enrich the lives of captive penguins:

For fish eating species that are active and capable swimmers, a live fish dispenser such as the one detailed in Chapter 13 would be a logical first step. This could be combined rather inexpensively with a method for detecting penguins' entry into the water, thus ensuring that fish continued to be dispensed only as long as the birds continued to show interest. Some institutions insist that live fish are not only costly but that they present health hazards not encountered with frozen fish. We have found this not to be the case as long as the live fish were bred within the institution or by a reputable source. In fact, we found live fish to be more germ-free at the time of delivery to predators than dead ones that had been defrosted and handled in typical manners (Markowitz 1982).

A second idea is to incorporate a wave machine such as I have suggested for marine mammals. With sufficient budgets, this could be integrated into an exhibit that encourages behavior typical of that seen in nature, such as using wave action in moving from water to rocks or ice flows. For many penguin species, an active current of water for aquatic exercise is imperative for maintaining health.

## Bats and Moths

Since I see the modern mission for zoos to be conservation education, the brief suggestions for enrichment and integrated graphics that follow are especially attractive to me.

If you were to ask students of behavioral ecology for examples of co-evolution of adaptations, they would almost certainly include echolocation used by bats to pursue moths and response mechanisms that have evolved in the prey species. The sensory adaptations of bats and some of their prey have long been intensively studied (Griffin et al 1960; Roeder 1965, 1970).

Current knowledge in this area has been used in a surprisingly limited number of graphics about bats in zoos, and has not been applied systematically in exhibit design or captive enrichment efforts.

Noctuid moths (Family Noctuidae) have one ear on each side of their thoraxes. This simple ear consists of a very thin sheet of cuticle attached to two nerve cells which have been labeled by scientists as "A1 and A2" sensory receptors. The evidence reported to date suggests that this ultra simple sensory organ is sufficient to trigger evasive maneuvers that frequently result in the moths eluding capture by bats using echolocation in pursuit.

Here are a few suggestions for ways to bring excitement and knowledge about co-evolution to zoo goers young and old while providing more stimulating environments for bats:

1. Visitor labyrinth graphics and games

Because the average zoo visitor has to be stimulated to pay real attention to graphics, I would suggest the use of two kinds of devices for visitor entertainment. These devices could be housed in the darkened labyrinth entrance that most zoos use to allow partial dark adaptation of the visual system, thus enhancing viewing of animals in nocturnal exhibits which are dimly lit.

A relatively inexpensive dynamic graphic that visitors can initiate by pressing a button could be used to illustrate the relationship between the known responses of both predator and prey when bats pursue moths. This could be accomplished either by using successively illuminated graphic components or a recorded video presentation. The graphic sequence might start with a dynamic visual representation of sonar emanations from a flying bat in pursuit of a moth. This, in turn, could be followed by illuminated diagrams illustrating how responses of the moth A1 and A2 receptors trigger different evasive responses depending on factors such as intensity and frequency of sonar emanations from pursuing bats. Dynamic graphics that visitors can initiate typically help to overcome the disinclination by many who come to zoos to read graphics at all.

Since decent dark adaptation takes considerable time, a second attraction for visitors waiting in the labyrinth area might involve interactive simulation games. I think that children of *all* ages, including me, might be attracted by an electronic two-participant game with one person playing the part of the bat and the other the role of the prey. The "bats" could generate echolocation signals and try to keep the signals focused on moths in order to use the echoed information to do their best to pursue and capture moths. Simultaneously, the "moths" could watch the screen closely for responses in their A1 and A2 cells and use the responses of the cells to the echolocation signals in trying to activate the appropriate evasive strategy. This would encourage returning to carefully study the graphic encountered earlier in the labyrinth because inappropriate responses by either participant to sensory stimuli would prevent winning the contest. No one wants to be a bat that can't catch moths or a moth that can't evade bats!

2. Enrichment for existing exhibits with sufficient space for short-range active pursuit

It is always fun to engage people with different kinds of biological expertise in the design of enrichment for an exhibit. I would begin by seeking the help of an expert in breeding of noctuid moths and others who are expert in the ecology of bats and moths upon which bats prey. Initial brainstorming would focus on the feasibility of finding a way to breed moths in sufficient numbers across all seasons to serve as prey for bats. Assuming that this was deemed reasonably possible, we would next focus on identifying typical circumstances where bats might be expected to search for prey and how to simulate these conditions.

Since the premise is that few, if any, zoo exhibits have sufficient space for long range active pursuit of moths by bats, this would require the occasional release of moths into an area near one end of the enclosure where bats ordinarily spend their time. Without having experimented on this approach, I do not know whether it can be successfully used to produce a decent exhibition of live predator and prey activities. If it was necessary to identify an alternative, I would use artificial prey, such as flying lures, which if touched by a pursuing bat would trigger the release of some attractive food. The shortcoming of this alternative is that visitors could not witness naturally triggered evasive behavior in moths. At best, well-designed equipment could make the path of the artificial prey vary in ways that simulated natural responses by moths when pursued by bats. Truly adventurous designers might wish to include a method by which visitors could occasionally trigger the release of real or artificial flying prey to a bat exhibit.

3. Big budget idea

Make a really lengthy captive bat cave with some twists and turns and a one-way window that allows visitors to observe bats and moths without the sight of visitors disturbing these animals. Work with a group of experts to develop a captive breeding program to ensure a healthy supply of the predator and prey species. Make necessary changes in the environment as needed to encourage natural predation and allow possible avoidance of capture by prey. There are others better prepared than me to design an exhibit such as this for bats and moths, so I will not make specific suggestions for construction and husbandry. Although I think the development of an extensive habitat like this would be a great idea from both educational and enrichment standpoints, it would require finding affluent supporters to champion and fund the venture.

If funding for natural science ever returns to high levels, it might be possible to produce an exhibit such as this, supported by a combination of zoo funds and research grants for further studies of echolocation and co-evolution in bats.

## Raptors

Many raptor rehabilitation centers are engaged in efforts to reintroduce injured animals to the wild after restoring them to sufficient health. A number of these facilities also strive to breed raptor species that are endangered or threatened with extinction, and to identify ways to enhance the probability of survival of fledglings. During time I have spent with my students and other friends who work in some of these efforts housed in zoos, I have always been impressed with the understanding and concern about the plight of these amazing birds evinced by people engaged in these efforts. They work tirelessly to find ways to avoid excessive human contact with the raptors during rehab efforts in the knowledge that becoming dependent on humans to provide care and food will make birds of prey much less likely candidates for successful reintroduction.

Much information is shared through networks disseminating useful information for working on behalf of raptors. This includes information about which species will only readily consume live prey; ways to estimate when individual birds are ready for attempts at reintroduction; descriptions of best medical procedures; and many other such findings of critical importance for successful rehabilitation, breeding, and reintroduction. These praiseworthy efforts could be bolstered if more zoos provided predatory opportunities for raptors that were as natural as possible in captive habitats. Carefully conducted research efforts are needed to determine the kind of zoo exhibit environmental features that will generate the most excitement for captive raptors and stimulate them to use their predatory skills. These studies could be followed by investigating the differences in attracting zoo visitors to spend time studying the raptor's behavior in exhibits where these birds have no opportunity to pursue prey, and those where prey hunting activity can be seen. Exhibiting the wonderful capabilities of raptors in pursuing prey could be a vital component in zoo efforts to generate support for conservation efforts.

Carl Cheney and his students spent considerable research time in determining critical factors that influence prey selection and capture by some raptor species. Even though a chapter he wrote titled "Predator-Prey Interactions" is now more than forty years old (Cheney 1978), the summaries of his research and pertinent studies by other investigators remain useful resources for the design of enrichment procedures to enhance the lives of captive raptors. Many of my brief suggestions for raptor enrichment are based on some of Carl's findings.

It has been established that with some effort it is possible to get raptors to feed on white mice as prey, and that rodents contrasting sharply in color with the substrate are more likely to be captured. However, a careful set of studies (Ruggerio 1975) determined that ferruginous hawks (*Buteo regalis*) initially find white mice undesirable, preferring prey that are of shades more like those rodents typically encountered in nature. If repeated presentation of mice is made, the hawks will eventually become less wary and choose to strike white mice. However, it might be a lot better for those planning enrichment to select prey more similar in appearance to natural prey for the species. This would be especially desirable in reintroduction efforts where the hope is to make the transition back to nature less dangerous. Being wary of odd prey is one way of reducing danger for predators.

While it would be stimulating for captive raptors if terrestrial prey were automatically released from time to time, it would not be so pleasant for humans tending them. It is almost impossible to devise humane methods that guarantee that rodents or other ground prey will not escape from a flight cage and infest the captive facility.

One method worth evaluating to see whether it increases raptor activity when these birds are on public display is the use of artificial prey. I would approach this very carefully by trying an inexpensive preliminary test, such as pulling faux rodents through a transparent tube (see Chapter 3) and seeing whether raptors would swoop toward the tube. I would reward any such behavior by the raptors by immediately tossing them a morsel of their favorite diet. Raptors have fabulously keen long distance vision so it is unclear to me, without testing, whether artificial prey running in a transparent tube and seen at a distance would be effective in triggering predatory behavior by these birds.

If this artificial prey method produced promising results, I would progress to a slightly more elaborate procedure by modifying the apparatus so that a food treat could be attached to the artificial prey to see if the raptor would strike at the artificial prey and consume the treat.

Here is the procedure I would use to modify the apparatus. Dado an opening the entire length of the transparent tube to produce a slot in which a ¼ inch nylon rod can easily slide. Attach the rod securely to the artificial prey, and ensure that the rod slides easily in the slot and maintains alignment so that it protrudes just sufficiently above the tube to attach a small portion of meat. This design requires securely anchoring the transparent tubing and ensuring that the rod protruding through the slot cannot be yanked out of the tubing by a raptor. One method to accomplish this would be to begin with a length of ½" nylon rod, and to turn one end of it down to a ¼ inch in diameter for a length of approximately ¾ inch. In order to minimize difficulties in pulling the artificial prey through the tubing, the prey item would have to run in a simple guide track installed in the tubing. This could be done by passing the ½ inch diameter rod through the artificial prey to run in a track at the lowest part of the tube. The length of the rod should be carefully measured so that the bottom of the rod will not rise above the guide track. The ½ inch diameter part of the rod should extend from the track to the narrow slot where the ¼ inch diameter section emerges.

I have not included a diagram of the proposed apparatus, because I feel sure that each craftsperson will choose their own favorite method to run the prey through the tube in a non-binding manner. Summarizing the essential idea for utilizing this preliminary testing device:

1. The protruding part of the rod is used to secure some favorite raptor treat.

2. The humans who will be pulling the artificial prey to "run" through the tube are concealed from the raptor's sight and remain silent throughout the procedure.

3. A treat is secured on the protruding rod, and the prey is pulled through the tube to the other side at unpredictable intervals .

I repeat the caveat that the above procedure *has not* been tested for effectiveness in stimulating raptor activity, and I cannot assure you that it will work. But, I do think it is worth a try. If it does work well, the apparatus could eventually be automated, but there are challenges in designing devices to automatically attach meat portions to the rod following the occasions when the raptor consumes the treat.

The idea below using aquatic prey is based on observations that I have made of raptors both in the wild and captivity who will strike at live fish that have been introduced into waterways. Thus I feel more secure about recommending it without reservation.

Where raptors that naturally prey upon fish are maintained in really large flight cages, my idea is to use a live fish dispenser such as that described in chapter 13. One possibility is to construct the cage in an area where a stream naturally runs through the enclosure. and use the feeder to periodically stock the stream with fish. In many cases it might be more efficient and less costly in terms of fish and maintenance to make an artificial waterway with a current that keeps water circulating into the cage. Where public exhibits are involved, architects could design this to appear quite naturalistic by making the shape of the "stream" irregular and incorporating rocks and foliage. The fish dispenser could be located in a hidden area in a part of the waterway that extends outside the cage. Fish species to be dispensed can be selected based on natural preferences for raptors. This is especially important when the hope is to someday reintroduce some of the raptors to nature.

Those working in facilities that use falconry methods to demonstrate the abilities of raptors might find Tim Gamblin's article about enrichment and humane ways to deal with raptor husbandry problems of interest (Gamblin 2004). I particularly like the practicality of some of his ideas and his recognition that enrichment for animals in his care requires tailoring to individuals within species.

Much useful information can also be acquired by visiting one of the raptor rehabilitation stations or accessing the information that they provide online concerning their efforts in healing birds and trying to ensure their fitness for reintroduction.

## Anteaters

A student in a course that I offered on the behavior of captive wild animals wrote an excellent lab paper that inspired me years later to include this brief section. The student was Manami Hayashi, and the proposed enrichment project was for Giant Anteaters (*Myrmecophaga tridactyla*) at the San Francisco Zoo. I have incorporated some of the valuable information collected by Manami in conceiving my suggestions for enrichment presented below.

In the wild, the primary diet for this species consists of ants and termites. Giant Anteaters have special adaptations for hunting underground colonies of insects. They depend on a well developed sense of smell for detection of these colonies. Strong claws are used to dig up the earth, and the long tongue is inserted into the nest to pick up multiple insects at a time. All of this is accomplished quite swiftly with the anteater spending only ten to thirty seconds at each nest (Redford, 1985).

Free ranging Giant Anteaters occupy relatively large territories and are solitary animals except for females with nursing young (Shaw et al, 1987). They employ fixed foraging routes which they seldom alter. It has been hypothesized that this allows them to feed on each insect colony at periodic intervals, thus avoiding overexploitation of prey (Redford, 1985). Since few captive facilities can provide areas for these anteaters the size of their natural territories, a major challenge in planning naturalistic enrichment is to find effective means to present attractive hunting opportunities with regularly provisioned nests along simulated foraging routes.

Because it is virtually impossible for most captive facilities to provide the 30,000 ants per day these mammals are estimated to consume in the wild, special diets are universally used. The following recipes require that the food used be in a form that allows small portions to be collectible by the anteater. It has been established that even very simple special efforts, like adding sandboxes for Giant Anteaters will result in considerable digging and bring apparent comfort to these animals (e.g., Leviele et al 1993).

### Recipe 1. Inexpensive and Primarily to Enhance Visitor Education

1. In a part of the enclosure easily observed by visitors, install a long 6 inch diameter PVC pipe horizontally near the perimeter of the exhibit. The pipe should be as long as is reasonably possible in the available space. Leave the ends of the tube open.

2. Secure the tube firmly in the most practical manner based on the substrate of the enclosure. If there is a dirt surface, it may be best to dig a trench that allows all but the topmost part of the pipe to be underground. One end must be accessible to the husbandry staff for easy loading and cleaning as described below.

3. Drill a large number of access holes in one side of the pipe, and be certain to smooth off the edges of these holes so that they will not damage the tongue of the anteater.

4. Make a long simple tool for placing food inside the pipe. This can be done by cutting a length of smaller diameter PVC pipe and using this smaller pipe to place food in the larger one.

5. Load some semi-randomly chosen access holes with food. Loading can be done in a number of ways, depending on the type of food that is used and the tool that has been devised. For example, if the food is granular you might use a rod as a plunger within the small tubing to dispense a small portion of food at each selected location. The same might be done with prepared meats or other moist foods by using a tight fitting plunger. In any case, regular careful cleaning of the food dispensing tool will be necessary between feedings.

The general idea for use of this proposed device is to partially simulate natural hunting activities by having the anteater find the food by searching, sniffing and digging, and then collect it by tongue.

**Recipe 2. Much More Time-Consuming for Husbandry Staff but More Naturalistic**

This requires either an enclosure with a dirt substrate or the use of a large number of sturdy boxes filled with soil.

1. Lay out a circuitous "foraging route" as lengthy as possible within the exhibit. If the substrate is dirt, the sites for hunting can be easily changed by using a tool such as a posthole digger to make holes large enough for food containers to be inserted in randomly selected places along the foraging route. If soil-filled boxes are used, you can scoop out enough dirt to place food containers in each box that you have chosen to contain any food containers for the anteaters to uncover.

2. For food containers, use 3 inch PVC caps for the container bottoms, and 3 ½ inch PVC caps chosen to fit loosely over the open ends of the 3 inch PVC caps to serve as easily removable lids. These containers will suffice to keep most dirt out and yet be easy for the anteaters to pull open after they use their claws to dig access to the containers. The PVC caps are easy to clean which is an advantage in using them as food containers.

3. While the resident animal(s) remain safely restrained, place food containers in the locations you have selected. Remove sufficient dirt to allow you to insert the containers so that the container lids can be hidden a couple of inches below the surface of the dirt after you have loaded the food.

4. Fill each 3 inch cap with a portion of food. Carefully insert this cap in the center of the hole that has been created for it, and firmly pat dirt around the bottom half of the cap to hold it in place. Set the 3 ½' cap on top, and then carefully complete covering the food container with dirt.

This recipe is designed to let anteaters exercise their abilities to identify food locations, and then dig to gain access to gather the food with their tongues. You can make this enrichment approach natural behavioral circumstances more closely by incorporating odors associated with food for the hunt. If you use food that has sufficient smell, you might add to the lid of your container a small "straw" that just reaches the top of the dirt to free odors sufficiently to allow anteaters to sniff out the food location. Alternatively, you could simply use some odiferous liquid, such as one that approximates the smell of ants, to place on the soil in locations above each hidden "nest."

I have considerable confidence that this recipe, though it involves frequent chores for husbandry staff or volunteers, will be sufficiently rewarding in its results to justify the effort involved. By presenting more naturalistic stimulation for the anteaters, this approach will lend itself well to attractive graphics and conservation education efforts that enhance the mission of zoos and wild life parks.

## Meerkats

In the same lab section in my course as Manami, Joshua Curmi produced some interesting ideas for Slender-tailed Meerkats (*Suricata suricatta*). These animals are a popular species with zoo visitors, perhaps in part because of the simian-like facial expressions they sometimes exhibit. The startled looks and erect postures they display when resting on their tails and hind legs to stretch the rest of their bodies into an erect posture are a surprise for first time visitors.

These interesting members of the mongoose family have a natural habitat extending through dry open country in southwestern parts of the African continent. Special adaptations for this environment include strong non-retractable claws and ability to dig up as much as their own body weight in clumps of earth to uncover insects and other invertebrate and larval forms found underground. In the wild they occasionally eat eggs, small snakes, birds, lizards and mice.

The recipes that I have just proposed for anteaters would also work well for this species, although I suggest a few changes to make things more naturalistic for meerkats. The artificial foraging equipment should be installed further beneath the surface of the earth. Hunting by meerkats often involves deep exploration for prey that they detect with their advanced olfactory powers. When using an enrichment treat such as mealworms, you can simply make deep holes, insert a batch of the treat in each hole and pack dirt in to fill the hole.

The additional recipes I offer here are only possible for those captive animal facilities whose employees and visitors are tolerant of witnessing animals capturing living mammalian prey. They also require a means for staff to recapture mice that elude the meerkats and leave the enrichment device. I wish that I was clever enough to devise a perfect institutionally acceptable method by which the mice had a greater "fighting chance" to escape and remain free, as they might in the wild. However, I cannot devise one that is practical and would cost any reasonable amount to implement. The second recipe comes closer to this goal than the first.

### Recipe 1. Energizing Alertness and Pursuit Activity in Meerkats

It is fairly easy to maintain a small breeding colony of mice, and to have a ready supply of potential prey for captive predators. There are also abundant commercial suppliers of live mice of many suitable strains. This recipe involves occasionally releasing mice for potential capture by meerkats.

1. Obtain some sturdy transparent rigid plastic tubing with an inside diameter of at least 2½ inches and at least 8 feet long. At one end of the tubing, cut away half of a one foot section of one side of the tube, so it will be possible for meerkats to reach into this open section.

2. Select a location near a side of the cage where husbandry staff can easily both introduce a mouse to the transparent tube and bait the open segment at the far end of the tube with some mouse treat. Imbed the tubing in the substrate in a manner allowing most of the tubing to be visible from the surface. Be certain to extend the portion of the tube where treats to lure mice are to be placed out of the reach of meerkats. Otherwise the meerkats may choose to eat the mouse treats.

3. At random times, bait the very end of the open segment of the tubing with food that is attractive to mice. Release a mouse into the other end of the tube and, close the end of the tube behind the potential live prey.

Although I have seen similar off-exhibit techniques for generating pursuit activity and learning to capture live prey work, I have not pre-tested this recipe. Consequently, there are at least two important factors to consider should you choose to give it a try. The first is whether mice which are not captured the first time they cross the open area to obtain food will choose to remain in the tube or scramble out and require recapture should they continue to elude the resident predators.

A second concern is that some mice may be wary, especially if clever meerkats use the loading of a mouse into the tube as a signal to hover over the open tube area. If this problem is encountered, there are a few possible solutions. One is to be certain to use a strain of mice that have been selected to be aggressive in their pursuit of food. Another possibility is to simply be patient and wait until a hungry mouse has no choice but to go for the food. I am quite sympathetic to the fact that this unpleasant situation for the mouse may preclude this as a method of enrichment for many readers. My own feeling is that I would prefer to use this procedure only in cases where preservation of species-typical behavior and vigor of meerkats outweighs the well being of captive bred mice. The fact that the mouse may not always immediately race to the open food end of the transparent tube might actually better simulate events in nature and lessen the tendency for meerkats to use introduction of a mouse to the tube as a cue to rush to the open end.

**Recipe 2. More Expensive and More Naturalistic for Both Predator and Prey**

I hope that some ambitious enrichment designer who works in an institution with policies that permit it will try this recipe. It comes closer to giving a fair chance for prey, and has several advantages for producing more typical predatory behavior seen in free ranging meerkats. A large captive environment is a prerequisite for effective accomplishment.

In the wild meerkats are aggressively territorial and tend to spread apart in foraging areas. It is not surprising that meerkats are notoriously combative when food treats are presented in captivity. Although I have not had opportunity to test this recipe, I have confidence it will lead to establishment of spaced hunting domains. This should decrease aggression and attendant injuries in meerkats.

1. Establish a breeding colony of mice.

2. In the home quarters for the mice, arrange a means for them to enter the open end of a transparent plastic tube in order to hunt for food.

3. Using strong rigid transparent tubing of at least 2 ½ inches inside diameter, make a very large warren-like maze of tubing firmly secured to the substrate with at least the top half above ground. The more extensive and wandering your tubing is arranged, the better.

4. At random places at least a few meters apart, make smooth holes in the tubing large enough so that a meerkat can reach in and attempt to capture a mouse that is in transit in the tubing. These holes can also be used by husbandry staff to randomly insert food treats for the mice by dropping treats through the holes and using a stick to push the mouse food out of reach.

I believe that if neither predators nor prey are overfed (as is the unfortunate practice in some captive facilities), the following behavior will occur. Hungry mice will eventually explore the tube in search of odiferous treats. Meerkats will sometimes observe the mice moving through the tube and attempt to capture them in places where it is possible to do so. Eventually, meerkats will establish some individual areas that represent their defensible location for preying on mice. Mice that are wary and fast enough to escape capture will eventually return to the home colony as they might in nature.

In the right circumstance, this recipe could be used to provide significant educational and research opportunities. If employed in a zoo or theme park, it might be augmented with a means to remotely open and close the entrance for mice to the tubing. This would allow a greater chance for maximizing public education about both species involved. However, I personally prefer letting the mice have access available at all times so that they can pursue food whenever they choose to do so. It would be interesting for observers to see whether the diurnal behavior of meerkats is modified by mice which do some food searching at night.

Whenever I write a new recipe, I simultaneously think about all of the other species for which this might be an effective technique. I think that this recipe for meerkat enrichment might work very well for captive raptors, especially those being prepared for reintroduction to the wild.

Readers who do not find it acceptable to use live prey in captivity, or who work in facilities where it is impractical to do so, might wish to try the simulated prey pursuit technique that we have successfully employed with felines (Chapter 3).

**A closing note**

Conserving diversity of life on our planet, and educating people about the need to work towards reducing overpopulation often involves morally difficult choices. I recognize clearly that many readers will identify some apparent contradictions between various personal values and ideas that I have expressed throughout this book. Nowhere is that more apparent than in some of the recipes in the current chapter. In chapter 17, I extolled the virtues of rats and mice, and why I thought they would benefit from enrichment...while in the present chapter I propose using mice as prey in institutions that wish to enhance public appreciation and conservation of other species. I have feelings of compassion for the prey animals that would be captured and consumed when people use those of my designs that include opportunities for predators to capture live prey. I believe that efforts in zoos, wildlife parks, and aquariums to improve conservation education, and to prepare captive members of endangered species for reintroduction to natural domains, will substantially benefit from the use of captive live prey. These efforts will likely be most effective if the prey emerge in a manner that simulates the way that this occurs in natural habitats.

# 20

# FORMS OF ENRICHMENT AND CONSTRUCTION OF DEVICES

## Choice of Type of Enrichment

The first thing that I always consider in designing environmental enrichment is the possibility of integrating some aspects of the natural environment into the captive setting to see whether this might enhance the lives of the resident animals. Some of the constraints that prevent zoos from being truly natural have already been discussed, but there are other concerns as well with respect to the perceived mission of the institution. An example that speaks to the point involves efforts by some of us to talk management of the San Francisco Zoo into safely enclosing a part of the forest within the zoo and making this a habitat for orangutans. Before we began our lobbying, we tried to foresee as many potential problems as we could in order to anticipate and counter any objections that might be made. We also began work on the design of environmental qualities that might reduce those immediate problems that we could foresee.

In thinking about these issues, we all had in mind the rapidly disappearing natural habitat for orangs, and that part of the job of the zoo was to protect and conserve this endangered species. Should anything untoward happen to an orangutan as a function of zoo husbandry, one could anticipate serious negative consequences and criticism of the zoo.

Keepers and docents had obtained permission to begin enrichment activities for some young orangs by transporting them from their exhibit to a forested part of the zoo grounds (figure 20-1).

Figure 20-1. Orangutans traveling in style

Because the young apes had never been exposed to real trees that were accessible, they did not immediately scramble to use them. Sturdy ropes were placed between some trees and as an enticement the keepers and some volunteers would vigorously swing the youngsters on these ropes. When my schedule allowed, I joined them in these activities. One day I climbed up into a tree and beckoned one of these orangs that I had befriended to join me (figure 14-2). After an initial show of caution, he became much more comfortable than me in the tree climbing trade and showed his pride by biting through my tennis shoe. But we *were* friends and he looked at me apologetically, lifted my tee shirt, and sucked my navel to show that he was sorry. What are friends for???

Figure 20-2. Two Apes Hanging Out in a Tree

Some of the curators were happy with this progress and now convinced that all we needed to figure out was how to properly enclose a large forest section in order to protect the apes from the public and vice versa. But I remained concerned about effects on the species conservation duties of the zoo, and how this enriched environment might bear on them. In the wild apes rarely but occasionally fall from broken limbs of trees and seriously injure themselves. Taking mature orangutans that had never lived outside a concrete jungle and introducing them to a truly naturalistic exhibit without the chance for them to learn how to deal with a forest was certain to raise problems. I felt that it would be better to let them learn how to regain their balance from accidental limb breakage while they were still young and resilient.

These young apes had a survival advantage that nature does not provide, receiving routine health exams and care when needed. So, my recommendations to the zoo were that older animals should not be introduced to the new enclosure. Instead, efforts should be made to provide them a softer substrate than the concrete in their existing enclosure in case they should fall from the steel and wooden pole climbing apparatus that had long been provided for them. The newly proposed forest exhibit should be the home for younger orangutans who could "grow into it."

There remained a number of problems to address. One was the question about whether there was a practical way to still have the young orangs visit with, and learn from, their adult conspecifics.

Advance plans were needed for assisting the veterinary staff in providing routine care. I suggested that we could provide the apes some irresistible treats that they could only get by entering a part of the forest where they could painlessly be restrained for exams and health care. Some ideas for this have been discussed in earlier chapters on primate enrichment. My point here is to emphasize that it is not always possible to simply introduce animals to a cross section of nature and expect that nature will take care of things. Often I have heard from zoo husbandry and veterinary staff that new spacious and complex exhibits have made their work nearly impossible. Proper anticipation of these difficulties and provision of prearranged solutions may greatly reduce such difficulties.

My personal feelings about the added risks that are inevitable for animals in a naturalistic exhibit are easily stated but not universally shared. If the more natural environment clearly enriches their lives, even if we are speaking about the last members of their species, quality of life for individuals matters most. Keeping clearly sentient animals like orangs largely powerless in excessively confining spaces as "gene repositories" is inhumane. I respect the fact that many respected colleagues have well-considered reasons for disagreeing with my position on this matter. It is not easy to decide whether enriching the lives of individuals outweighs the importance of conservation of a species.

**Combining nature with captive needs**

In planning enrichment for tigers and gibbons in a large Sumatran rainforest exhibit in the Panaewa Rainforest Zoo which was under construction in Hilo Hawaii (Markowitz, 1982), I had to consider several critical factors. First this exhibit was *not* in Sumatra and not all wildlife and plant forms indigenous to that area would be present. Second, we had to accommodate husbandry and health care requirements. Since one of the resident species was to be tigers, the zoo staff wanted to be certain that they would still be able to use food to lure them into night quarters where they could be cared for safely. Additionally, there was the ever present concern that there were regular occurrences in Sumatra that the public and regulatory authorities would not condone in zoos, such as the capture and consumption of other residents of the area by tigers. Still, I did not want this to be an exhibit that looked natural but did nothing to encourage naturalistic opportunities for the animals, or failed to capture the attention of zoo visitors long enough for them to learn firsthand about the beauty of tiger and gibbon behavior.

I had spoken in great detail about my concepts with Jim Juvik, who had taken leave from his professorial work at the University of Hawaii in Hilo to do planning for the zoo, and act as the initial director. Jim and I did initial planning for what we would use to make the exhibit, which was in the first stages of construction, a stimulating one for the resident animals, and attractive for zoo visitors. I drew some plans for how we would accomplish this, including the devices to be used in our behavioral enrichment efforts. Ron Fial and I designed details of construction for each device, and along with our trusty enrichment team began construction of artificial prey for the tigers, a system to encourage foraging activities by the gibbons, and means to encourage the public to learn about the natural abilities of the animals. We also provided an opportunity for visitors to safely participate in varying the enrichment activities for the tigers.

Although I always prefer to provide fully natural opportunities for captive animals, there are some advantages from the standpoint of the education mission of a zoo or wildlife park when you are forced to provide more limited opportunities such as artificial prey "capture." For example, since the movement patterns of artificial prey are known in advance, visitors can be informed about where to look to see naturalistic behavior by predators.

With no training to do so, the tigers in this rainforest exhibit actually hid and then stalked the prey before they pounced on it, perhaps in part because they could not be certain which prey would currently become active. People visiting nature reserves rather seldom have the opportunity to directly witness predators succeed in prey capture. The artificial prey aspect of this exhibit provided a higher probability of observing the behavioral beauty of a predator in action than is typically the case in the wild.

Of course, the problem with ambitious work such as this is that it requires substantial budgets for planning and construction and routine equipment maintenance. Most often one needs to plan more modestly in enriching animal lives. This is especially true if you are not going to be present to supervise the daily use and maintenance of enrichment devices.

Unforeseen budgetary problems for the local government resulted in replacing much of the zoo staff with lower paid employees, and they could not honor the written agreement to hire a skilled person to maintain the equipment and supervise daily enrichment efforts. Despite the crowds that came to see the animals actively engaging in activities that demonstrated their natural abilities, this project was short-lived. Our enrichment team returned to their mainland duties. Students who had volunteered to look after the enrichment equipment during the summer session returned to their university studies. There was no one available to see to the use of the equipment and carry out routine chores like loading food in enrichment devices, and Jim Juvik returned full time to his professorial position. Despite its popularity with resident animals and visitors and the positive media coverage of the enriched exhibit, the equipment we had worked so hard to develop and install went into disuse.

**About Those From Whom You Need Help in Your Enrichment Efforts**

Nick Lee is one of the finest people, and best coworkers whom I have ever had work with me on environmental enrichment. He began as a volunteer who had heard about our work in the Portland Zoo and asked if he could be of help. Soon it became clear that Nick was doing more work, and accomplishing it better, than some enrichment team members who were paid for their efforts. The only sensible thing for me to do was to put him on the payroll. Nick worked with us for all the rest of the remaining years in which I directed the Oregon Zoological Research Center. There are countless others who have voluntarily provided immeasurable amounts of help in my enrichment efforts. This includes zoo and aquarium personnel, laboratory workers, and other professionals involved in animal care and maintenance that stayed after hours or came in on days off to provide assistance as volunteers. I am also, as always, indebted to my wonderful family who tolerated the constant mess in our home that resulted from my design efforts and prototyping. Krista, Tim, and Jenny did everything from climbing inside and being the first animals to test new enrichment equipment, to weaving hemp ropes for primates to use. Jenny even learned how to do precision soldering to help build electronic devices that I designed. I mention these examples at the outset of this section, lest you conclude that I do not appreciate or welcome the help of volunteers.

I also have some fear that readers will find the following suggestions pedantic. But, I offer them anyway in the hope that they may save some of you from repeating some of the blunders that I have made in accepting help. Often these bad decisions have greatly delayed, or in the worst cases precluded, completion of enrichment projects. What I summarize here is based on more than forty years of occasional disappointment and regrouping efforts.

**Rule 1: Always Get a Signed Contract From Volunteers**

Each time that a new enrichment project receives some notice, I have been blessed with dozens of offers for voluntary help. For many years I faithfully and trustingly accepted these kind offers whenever it seemed that the help would be useful and that the offer appeared sincere. Then I had one of those private accountings that we all occasionally have with ourselves. I realized that among the previous several hundred commitments of help that had been offered, less than ten percent of the help had been delivered as promised. My accounting also shed light on the fact that I was spending much more time trying to get volunteers to meet their promises than the time that it would take me to accomplish the work myself.

*But, real help is always welcome,* and much more can be done to improve the lives of captive animals with good help. My solution was to devise contracts that have proven very useful in keeping some friendships and maintaining my sanity (albeit to a limited extent) and the sanity of those who work with me. Getting a signed contract and asking the volunteer to please keep their copy of the contract in a safe, accessible place has a number of benefits. Should you choose to adopt this method and it offends some potential helpers, you can always blame the requirement on me and let them read this chapter.

A well constructed signed contract has these advantages:

1) Leaves no chance for faulty remembrance of promised help and promised dates of that help.

2) Provides leverage for "firing" volunteers while still thanking them for their partial or promised help.

3) Reduces the time required to hear endless excuses about why paid employment, mysterious undocumented illness, etc. have prevented carrying out the work "yet."

4) If your work is conducted under the auspices of some institution catering to the public, having such a document will be of great value when you are inevitably sometimes accused of having been unappreciative of a particular volunteer.

5) Should you choose to extend the original contract deadlines, you will have increased leverage in having new ones met.

**Class Based Project Caveats**

I have taught courses on behavioral enrichment for more than thirty years. Frequently students elect to design enrichment requiring complex equipment. For many enthusiastic individuals, this is their first experience in design and construction of mechanical and/or electronic equipment. They approach some captive animal facility and seek permission to install the equipment once it is completed. When I suggest that a more modest approach for their first effort might be better, they are highly likely to tell me that there is some family member, significant other, or friend who has assured them it is easily accomplishable. The students feel secure because their volunteer helpers are self-proclaimed experts in building this kind of equipment. *Expert helpers* inevitably appear when the ideas that the students have come up with are attractive and inventive.

My "accounting" reminds me that of the many hundreds of offers from these *experts,* who promised students that they could design and construct equipment suitable for use in some particular institution, only *one* of the well meaning folks that verbally obligated themselves to students has *ever* completed the work. This kind and talented engineer/craftsman was the student's father and was able to spend all the time needed in his own workshop because he was retired!

This is a very special problem for those who are just getting started in enrichment efforts and have limited personal knowledge of engineering and construction techniques. The problem is magnified if the person offering the free help has less than adequate knowledge about the special requirements of equipment that can be used safely with animals. There are also special techniques in construction required in order for the equipment to last very long in environments where it will frequently be exposed to corrosive forces and accidental damage.

Failure to anticipate the kinds of problems described above can lead to sad consequences. A specific example may be helpful. A wonderful, bright student traveled 170 miles three times a week to take one of my enrichment classes. She was especially attracted to the fact that mobile artificial prey had proven to be exciting for captive predators, and chose to plan artificial prey to make life richer for some wolves in a local zoo. Since the zoo agreed to allow the effort but could provide no money, she prepared a grant proposal and brought me a draft of it. Her plan was to run artificial prey all around the perimeter of a very large exhibit, using a mechanism similar to that used in dog race tracks. The exhibit was built on a hillside with rolling hills and a fence specially built to follow the terrace-like perimeter. She asked about how much it would cost to design and build this equipment. I estimated at least a hundred thousand dollars.

Shaking her head in disbelief, she told me that her volunteer engineer, who was both experienced and employed in the design of apparatus, had said that he could do it for less than a thousand dollars! I gave her my pledge of undying praise and respect for her friendly engineer if this effort proved successful. It was entirely up to this bright student whether to proceed and convince the zoo that this was a safe and worthwhile venture. She was disappointed when I indicated that I could not ethically co-sign a grant proposal when I believed the proposed project budget was so grossly inadequate to accomplish the work.

The zoo was easier to convince, since there was no real cost to them and the potential for a free improvement for the animals was enticing. A zoo administrator co-signed the grant proposal thus satisfying the requirement of the funding agency and a small grant was successfully obtained based on the general concept. As part of the course requirement, the student brought the first real sketch of the equipment to me. I immediately noticed a number of serious potential problems inherent in the design. She anticipated that the artificial prey might occasionally stop moving because of changes that occurred in the exhibit as a function of debris or earth shifting in storms, or changes made by landscape crews. While instructions were to be provided for how to fix it to run again, there was no consideration of what might happen if a wolf was in full pursuit and close to the prey when it stopped suddenly.

I shared with this student a newspaper article about how a valuable greyhound had been near fatally injured when a "rabbit" stalled suddenly at a Phoenix racetrack. Another evident problem, which dismayed her when I pointed it out, was that there was nothing to prevent the wolves from injuring their paws by catching them in the track upon which the prey ran. When these concerns were presented to her engineering friend, he gave up on the project.

The student took a less ambitious plan for animated artificial prey for the wolves to *another* friend in the hope of spending the grant money in a way that might still benefit the wolves. This friend was cheerfully confident that he could do the work because his salary was paid by the university shop where he was employed to help students. Even though this was a design for installation of a short artificial prey run in a flat portion of the enclosure, making it safe for wolves still presented many challenges. After three years of waiting for the completed plan, this valiant student gave up and went to work full time on completing her doctorate. She topped the environmental enrichment class on all exams, did a great job on the final lab write up, and we both considered the failed project a good learning experience.

**Rule 2: Have Clear Written Deadlines for Those You Employ and (sort of) Enforce Them**

Much of the best of what we accomplished in the early stages of my work on behavioral enrichment could not have been done without the technical expertise, and equipment manufacturing skill of Ron Fial, and the dedicated and skilled routine equipment maintenance provided by Nick Lee. It has also been my pleasure to work with one or two architects and engineers who actually came close to meeting contractual deadlines. I mention this at the outset of this section lest you think that I am just a demanding and cranky person without respect for technical staff.

If you are independently wealthy or in control of purse strings that allow you to pay money for design and construction of enrichment equipment or exhibit improvement, you would be well advised to never trust the always optimistic time estimates of architects, engineers, and those who will build what you need. To corroborate this, speak with any zoo director. When you begin your own search you will doubtless be exposed to some wonderful preliminary enrichment designs so compelling that you dearly want them to come to fruition. Through painful experience I have developed the following guidelines that will partially reduce grief:

When a time estimate is provided to you, tell the professional with whom you contract to *double* their time estimate, and then execute a contract that binds them to meet this generous deadline or not be paid.

If they fail to meet the deadline and show nothing of value by then, find someone new. No matter how much you personally like the individual and the design, fire them if you reach the point where *six* times the contractual time has passed and the work is not fully completed.

Expect that whatever the original estimate given, it will be wrong by at least a factor of three so that you do not cry too much when you find this to be the case.

Enforce the KISS (Keep It Simple, Stupid) principle whenever you design apparatus in collaboration with engineers or technicians. Do not let designers incorporate "bells and whistles" that you did not ask for unless they are *so* attractive that you are willing to wait longer for apparatus to be completed and successfully tested. Be prepared to accept the fact that each of their proposed additions to your original plans will probably add to the potential service requirements for the equipment.

Of course these suggested guidelines are very arbitrary and idiosyncratic. But I stand by them as suggestions that will reduce your anxiety and maintain your sanity when dealing with the wonderful, creative, optimistic folks you may employ in your enrichment efforts.

**Some General Ideas About Available Equipment That May Serve in Enrichment Efforts**

In the recipes throughout chapters 3 through 19, I have made suggestions for the implementation of particular kinds of devices. I thought it might be useful, especially for those who have not spent much time looking at the technical end of enrichment efforts, to be introduced to a number of kinds of equipment. Of course each individual who might advise you on choice of components will have their own conception of what is best, and most technicians and engineers will be more up to date than I am with respect to available components and technical advances.

**Nutrition Delivery**

Ever since the appearance of the first animal life forms, there have been selective pressures favoring those able to gain the nutrition necessary for survival. There is a long evolutionary history that leads to the complex contemporary animals that are the focus of current enrichment efforts. It is not surprising that animals universally find sources of nutrition, and the chance to hunt or gather them, highly motivating. When one considers the seasonal, and famine-induced, reductions in food in natural habitats, it is also not surprising that the most successful individuals are likely to be those that constantly practice their abilities to obtain nutrition. This is, in my opinion, the reason that animals continue to work to obtain food even when they are not hungry. It is also the reason that we have found it unnecessary to deprive animals of food (unless they are unhealthy due to obesity) in order to get them to try new foraging opportunities. In fact, the exercise generated in actively hunting or gathering food in enriched environments helps animal fitness including weight regulation (Schmidt and Markowitz 1977).

One option is to use entirely passive forms of enrichment that simply allow the animal to work to gather food. We have successfully used some of the inexpensive foraging devices championed by Kathryn Bayne (Bayne et al, 1992). Artificial pelts are stretched on sturdy frames that can be safely attached in a place where animals can work to remove treats like sunflower seeds that have been imbedded in the pelts. Another frequently used passive technique employs objects such as hollow nylon balls filled with food that animals can jiggle out. While the use of passive devices is certainly better than providing no opportunity for animals to forage, our studies have demonstrated that there are some great advantages to providing responsive environmental components that allow for challenging and active gathering of nutrition (e.g., Markowitz & Line 1989).

Because the availability of electrical power at sites where reinforcement is to be delivered is often problematical, and the alternative of battery power requires monitoring of charge and/or routine time consuming battery replacement, it is always preferable if the need for such power for devices can be avoided. One alternative that may show increasing use in daylight environments is to use solar cell battery equipment. Another little-used means of delivery of reinforcement for active effort is to use the animal itself as the motive force.

An example of this is the foraging device which I originally designed for use by ground burrowing animals, and later modified for use in enriching the lives of pigs employed in research. A full recipe for making this kind of device for pigs appears in chapter 12. There are illustrated descriptions of more recent versions of animal-powered active foraging devices at the end of the current chapter.

Another possibility that I don't believe has yet been tried in enrichment efforts is to design equipment in which the animals foraging efforts are harnessed to provide sufficient electrical charge to deliver some nutritional treat. Animals foraging by turning a cylinder or operating a lever repeatedly could, for example, generate the necessary charge to activate a liquid solenoid that delivers measured nutrition. Animals that use running wheels can also potentially serve as dynamos to generate electricity with properly engineered equipment.

Good quality commercial electrically driven feeders such as pellet dispensers are both expensive and often unreliable in typical enrichment applications. When one is working in a laboratory setting and immediately alerted to malfunctions due to such things as food jamming a feeder, it may be possible to fix the problem quickly. But in settings such as zoos, where the feeder may not be attended to except for filling once daily, a jammed apparatus is more than a minor source of inconvenience. I have found that the only feeder that we could inexpensively build and count on to last for years without such malfunctions is a belt type feeder, the construction of which I described in chapter 3. The major disadvantage of a belt type feeder is that it requires a few minutes for the human to manually arrange food on the belt. This is more than compensated for by the fact that this type of feeder does not require uniform sized pieces of expensive factory prepared food. Feeder belts can be loaded with a wide variety of food providing a variety of attractive and nutritional foods for the resident animals (figure 20-5).

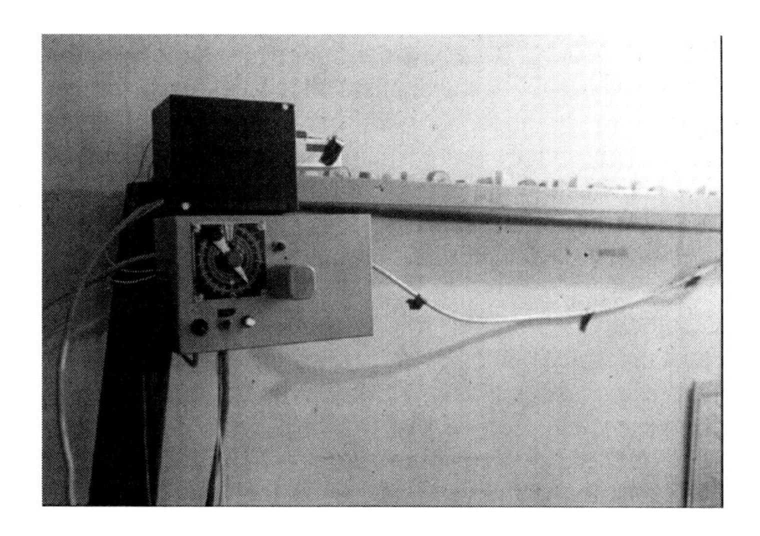

Figure 20-5. A belt holding a whole day's supply of various nutritious foods which are attractive for gibbons

If nutrition in liquid form is appropriate, relatively inexpensive excellent quality liquid valves are now available. These can easily be activated through timing devices that allow measured amounts of nutritive liquids to be delivered.

One last general statement about food delivery devices: If commercially available feeders are to be employed, there tend to be many fewer malfunctions by those that are motor driven as opposed to those which use solenoids with snap action. In addition to the fact that the latter are more likely to crush food thus jamming the mechanism, solenoids build up magnetic flux that eventually results in them sticking and being inoperable at unpredictable times.

There are other things besides food and water that can be provided to motivate animals in enrichment efforts, such as access to conspecifics, and even access to the animal's own reflected image or the image of others. In our enrichment research, we have found that these alternatives are not nearly as powerful rewards as food. There are, however, some circumstances in which they are especially useful because it is impractical to use nutritive rewards. An example is when animals are on a restricted dietary regimen such as in some research protocols. It is important in carefully evaluating potential usefulness of various reinforcement techniques to keep in mind that humans are not the only animals that show individual preferences. In one case, we found a rhesus macaque that loved the opportunity to turn on music while like-aged conspecifics showed much more limited inclination to take advantage of this opportunity.

Non-human megavertebrates also show intra-specific preferences for various nutritive resources much as we do. In a few cases in our research, we have been able to identify factors in life histories of individuals that have determined these preferences.

**Providing Active "Prey"**

While descriptions of the use of such devices may suggest that they are easy to build and maintain, animating reliable artificial prey systems is much more difficult to accomplish than it might at first appear. We have sometimes had to abandon several prototypes before settling on a device that was practical and provided long-term reliable service.

Aerial targets such as those representing flying birds can be constructed with reversible motor and pulley systems. However, this presents complex problems if they are to directly carry some form of nutrition which the predator can "snag." To allow continuous periodic availability of this kind of active prey capture opportunity, there must be a mechanism that feeds back information when capture has occurred, and the lure needs to be reloaded with food. The use of aerial devices is not highly recommended except in those cases where there is constant surveillance by husbandry staff or assigned volunteers. There is always the inherent danger that an animal may be accidentally "captured" by the "prey" resulting in injury to the very animal whose life you wish to enrich. An example would be a leaping feline catching its claw in an artificial bird.

Moving artificial prey on some sort of track or rotor is somewhat less fraught with problems, but there are some important cautions that must be observed. For example, there must be some relatively fail-safe method to ensure that the predator cannot be caught between the track and the artificial prey that runs on it. In his work at the Brookfield Zoo, Ron Snyder (Cheney, 1978) addressed this problem by guaranteeing that the predator could never capture the prey. He did this by requiring the predator to begin its effort from a static position that was too far from the track for the predator to actually come into physical contact with the prey before it disappeared out of site. To entertain zoo visitors, the predator then "pursued" the prey into an out-of-sight area where it could be provided prepackaged food while leaving the impression that it might have captured prey. I personally prefer systems that allow visitors to know exactly what is really going on, and not have to ask someone if the predator ever ate the prey.

It is possible to build systems that present very low risk of injury to the predator as a function of contact with artificial prey. Examples are the devices which Ron Fial and I designed more than a third of a century ago to animate a "squirrel" and a "rabbit" which periodically scampered across large mounds on the ground and were preyed upon by tigers in the Panaewa Rainforest Zoo (Markowitz, 1982). Among the safety features that prevented any mishaps throughout the use of this apparatus was the lack of any crevices in the carefully sculpted artificial prey. We were greatly appreciative that the sculptor's artistic sensibilities were not offended by having to eliminate any of the real life features that might have snagged the tigers' claws (fig 20-6). The other major precaution served to protect both tigers and prey. It involved a counter balancing system so that when the tigers successfully pounced on one of the artificial critters, the squirrel or rabbit would be knocked beneath the ground where they could cause no injury, and could not be mangled by the predator (figure 20-7). The public was then alerted that the tigers would be automatically receiving fresh flesh from a device directly beneath the major viewing pavilion.

After successful hunts the tigers raced toward the public and were delivered fresh refrigerated meat by means of belt type feeders.

Figure 20-6. Molded Prey Ready for Installation

Figure 20-7. Checking Counter Balancing System

It took the better part of a year to design and build this equipment and to prepare it for successful use in one of the rainiest parts of the U.S. But in final form, the appearance and safety of artificial prey paid us handsomely for the work (figure 20-8).

Figure 20-8. Final Installation Blends Artificial Prey Mound with Surrounding Area

In more recent work, we have found many advantages in using acoustic "prey" as described in earlier chapters. We were all somewhat surprised that the sound of prey species was often equally effective, or even *more* effective, than visual replicas of prey in encouraging pursuit by predators. Because of modern digital technology, sounds can be delivered reliably at a fraction of the cost of animated visual prey, and the use of sound eliminates the routine service necessary to maintain mechanical devices. Sounds can also be more "genuine" in the sense that real living prey sounds can be digitally recorded and then travel through systems of speakers to provide apparent movement of the prey through the environment. Since in the wild predators often depend on their keen sense of hearing to detect and stalk prey, pursuit of acoustical prey may also provide more naturalistic environmental enrichment. Unless one foolishly places the acoustic prey paths in areas presenting danger to the predators, most of the concerns that are apparent for mechanical prey are avoided with acoustic prey.

On the horizon there is the possibility of continued improvement in methods to deliver holographic representations into complex environments. One could then safely provide the animals with visible holograms of prey for them to pursue. Major concerns remain even when this technology becomes affordable and practical for such uses. Considerable evaluation will be necessary to ensure that exceptionally realistic appearing non-graspable prey are not a source of serious negative stress for the predators. Being certain that the dignity of the predator is not reduced in the eyes of observers requires careful planning. As Disney has taught us, this approach would undoubtedly be attractive to humans who come to be entertained. But it would be unseemly to have those occasions where predators failures in capture attempts make them seem like buffoons.

One of the real sources of joy and inspiration in planning for enriching the lives of captive animals is inventing your own system and deciding what will be most attractive and motivating for these animals. This may be accomplished in countless imaginable ways, but the safety of animals and the availability of time and money to accomplish ambitious enrichment plans must play paramount roles in your planning.

**Some Options for Programming and Control of Enrichment Devices**

Today, those planning complex behavioral enrichment have wide and relatively inexpensive options. Computers have become so affordable that they are often the best choice for interfacing all aspects of enrichment paradigms. Using a computer allows greater flexibility of design and ease of making improvements in the design without having to necessarily alter apparatus components.

One drawback of using typical computers of the home or office type is that they must be protected from the hostile environments often presented where enrichment is to be undertaken. One means of accomplishing this is to place the computer in a separate area such as a protected and appropriately temperature controlled part of the husbandry staff area. When budgets are not too tight, an option is to use industrial computers designed to be placed in controlled environment "shells" that provide protection from damage. Besides flexible programming, the other advantage of using even the simplest contemporary computer is that one computer can be used to control enrichment for several animal environments.

Of course thoughtful readers will recognize that should the computer "die," all of the associated equipment in all of these environments will stop operating simultaneously. If skilled technicians are available, an attractive alternative is to use one of the small programmable control units that have become popular in industry and robotics. These are essentially highly reliable minicomputers (sometimes based on a single chip containing thousands of components). Using a programming device to install instructions for operation, these units can be employed to register responses and control nearly any imaginable enrichment equipment. While the control devices themselves are very rugged and not too expensive, the programming equipment for these devices represents a one-time larger cost. Fortunately, some manufacturers have begun to market industrial controllers that can be programmed using a standard personal computer and an inexpensive interface.

In the case of severely limited budgets, all is not lost if one is willing and able to spend the time to make their own control system out of inexpensive integrated circuits (ICs). I have had students (prepared only with a firm prior grasp on principles of physics) produce usable enrichment controls for surprisingly complex paradigms for less than a hundred dollars. A good alternative for beginners wishing to take on this challenge is to use CMOS ICs which are quite cheap, very tolerant of minor power fluctuations, and for which there are all sorts of free manuals available from IC manufacturers. Many of my students have found that books such as Don Lancaster's CMOS Cookbook (1977), along with a little tutoring in how to handle and properly connect electronic devices, is all that they need to learn to design and construct simple electronically controlled enrichment equipment.

**Audio Chips**

Mass production of solid-state recording devices for telephony has driven the cost of chips that record sound to amazingly low prices while simultaneously allowing the development of increasingly sophisticated capabilities. This is indeed a blessing for those of us who have spent too much of our lives repairing and servicing tape recorders, wire recorders, compact disc recorders, etc. Older media storage and playback techniques may continue to be useful in household applications, but are no longer sensible alternatives for continuous use in enrichment applications.

Media degrading and mechanical breakdowns are no longer problems if you record sounds digitally in solid state memory which will probably last longer than any other components used in your efforts. If you have computer-based enrichment apparatus and sufficient memory space available, you can simply record the sounds directly in computer memory for playback. If you wish to incorporate sounds and have a simple enrichment design that does not require a computer, then recording on a chip is a reliable inexpensive alternative.

In addition to the production of acoustic prey, there are myriad other potential applications for audio chips in enrichment efforts. A few that immediately come to mind are: providing feedback to visitors about what is happening in enriched environments; providing soothing sounds such as the voices of familiar humans or sounds of conspecifics when animals must be left alone; providing a general ambience for exhibits (such as the sounds of audible critters that would be present in the natural habitat but do not appear in the captive location); and providing immediate feedback to hunting or gathering animals letting them know that their efforts have succeeded

Most major general distributors of integrated circuits now stock inexpensive digital recording chips. These come in a variety of forms, but in general the cost is a function of the maximum length of recording and the sound quality required. A typical inexpensive chip may be limited to recording only brief sounds if maximum fidelity is required. Cascading of these chips may be necessary if you wish to store long passages such as lengthy bird songs. Devices necessary to program chips with sounds are available in a number of different forms, ranging from expensive sophisticated manufactured programmers, to kits costing as little as twenty-five dollars. Some of these kits include all of the necessary components and instructions for assembling a programmer to which the user simply attaches a microphone and a potentiometer to adjust sound levels.

Even as I am doing final editing on this chapter, I have just discovered that a new generation of low cost digital recording chips will allow storage of much longer storage of audio segments, and commonly available computer software can be employed in storing the sounds on the chips directly from sounds that you have saved on your computer. It is also possible with a little investigation to find high quality miniature plug in computer drives costing less than twenty dollars. These drives have such high capacities that you can record a relatively infinite number of sounds that you might wish to use in enrichment applications.

Another significant benefit of using acoustic-based enrichment is that the energy level required to transmit sounds from areas that are not accessible to the animal to the speakers within the animal environment is low. In most cases, energy levels will introduce no hazard to resident animals. Thus none of the elaborate precautions necessary to protect animals from dangerous electrical currents are required, assuming proper design and installation.

## Brief Annotated List of Other Devices Useful in Improving Animal Lives

### Motion Detectors

For tight budgets, detecting the movement of animals and producing an output for enrichment equipment can be accomplished for very little cost in materials by modifying mass produced devices that are readily available at hardware stores. Infrared motion detector devices used to turn on lights when someone enters a driveway or other area surrounding a house are the “mother lode." They are very cheap, and a little inspection and ingenuity will allow you to identify units that include all of the components necessary to provide outputs to enrichment equipment devices and response recorders.

The majority of these detectors have controls: 1) for the user to limit the range and area in which large warm bodied animals trigger them; 2) to adjust the duration of “on” time; and 3) to adjust the sensitivity based on size of animals to be detected. Some of the slightly more expensive models have additional controls such as ones that allow the setting of minimum intervals between times that the apparatus can be triggered. It is a fairly easy task to remove the spotlight receptacles and use the signal that would ordinarily turn on illumination as an output to operate relays that initiate operation of enrichment equipment and response recording devices.

Another kind of easy-to-use detector device is the simple gyro-switch type device often employed to detect tampering or excessively coming into contact with unoccupied automobiles. The sealed switch mechanism itself is very simple and can be purchased locally at many electronics stores at low cost. We have found gyro-switches easier to use and more reliable than more costly switches that require mechanical operation and spring return mechanisms. We have used five dollar gyro-switches to detect when a large animal pounces on some part of a tree and used the output of the switch to control enrichment equipment. This method has produced trouble-free and very long lasting equipment.

There are all sorts of more sophisticated (and expensive) detectors such as those that use radio frequency fields to detect entry into an area and monitor movement. My experience has been that *the greater the complexity, the more often these devices are likely to malfunction*.

### Ways for Animals to Initiate Changes

Some of the equipment listed above can obviously be used to allow an animal’s action to trigger an environmental change. For example, we have used passive infrared detectors from driveway-type motion detectors to empower animals to initiate environmental changes by moving to a specific part of their environment. We have used this method to allow a number of species to control things like the onset of showers or mists and the sounds of prey by moving to a hunting area, etc.

For those applications where the animal is required to touch something to cause a change, the use of contact detectors (similar to those used in modern elevators) has proven to have great advantages over electromechanical switches. Contact detectors properly installed are much less likely to be damaged by animals, last longer because they do not have moving parts, and are frequently much easier for animals to learn to use than are switches that require significant pressure for operation. Good factory made contact detectors remain surprisingly expensive considering their simplicity. If budgets are tight, a simple contact detector can be built with CMOS and other inexpensive components for less than twenty dollars. However, one should carefully weigh factors, such as the time required to debug prototype equipment and to properly protect it from the elements and from damage by animals, versus relative cost of sturdy commercially produced detectors.

Another very reliable and inexpensive alternative for some manipulanda is the use of magnetic switches. These can be purchased for just a few dollars because they are mass produced for home use. Inexpensive combinations consist of two units both encased in protective molded casings: a permanent magnet which requires no electrical connection; and a reed relay with leads extending from each end of the casing. The relay is operated when the magnet passes in close proximity to it. The reader may be familiar with use of these switches in home security systems to detect the opening of windows. We have found this a reliable and inexpensive alternative for many enrichment applications. For example, after encountering many problems with corrosion and mechanical breakdown when using other kinds of switches to detect the movement of elements in a "xylophone" for dolphins and seals, we found that changing to magnetic switches eliminated these problems.

I am currently working on the design of some new enrichment equipment that uses versatile digital devices that have recently arisen in the consumer market. These include rugged touch screens that allow choice of many audio and visual materials including motion pictures, all of which can easily be stored in digital form on the device. Touch screen "tablets" increase in price as you choose higher storage capacity models, but even the least expensive will store many gigabytes of digital media, which is more than ample to store abundant materials for the captive animals whom we want to empower to be able to choose when and what they wish to see and/or hear. The latest generations of tablets use long-lasting high efficiency batteries that can go ten or more hours in use before recharging is required.

With proper programming of the devices, and well-designed enclosures to make them withstand the punishment that animals are likely to inflict, these devices can allow animals to touch displayed pictures of various choices, such as a picture of a human they know, thus initiating playback of a recording of that human saying soothing things in a familiar voice.

Things I like about using an electronic tablet as the heart of an enrichment device: 1) relatively modest cost for the device and components needed to make it safe for use by captive animals; 2) will allow a nearly endless variety of AV items to be added; 3) minimal time is required to design and build the entire apparatus; 4) when designed and manufactured correctly the enrichment equipment will require no routine maintenance save occasional charging of batteries; 5) ease of regularly changing the items available for selection by the animals in order to maintain their interest.

**Closed Circuit TV and Infrared Cameras**

For institutions that can afford such equipment, closed circuit television is especially useful in allowing husbandry staff and researchers to monitor the behavior of animals in a number of exhibits at once. This is very expensive if one employs video specialists to design and install slick systems. We have accomplished this for a fraction of the cost by searching through catalogs of brand new government surplus equipment and doing our own interface wiring. Low cost security systems with multiple television monitors and cameras have also become available, allowing those with moderate budgets to employ them in monitoring use of enrichment equipment. A significant advantage of using TV surveillance is that one can observe the behavior of animals without your presence affecting their actions.

Another possible application of television equipment is to use it to provide animals like great apes the opportunity to select television programs and to turn the television monitor *on and off.* I have found that the programs selected to watch by animals such as orangutans are often more interesting (and more intellectually stimulating), than the programs that are most frequently watched by some of my friends.

An important aspect of improving the well-being of animals in zoos is to provide them with privacy at appropriate intervals. An extreme example of this is the need to provide a den area for pregnant polar bears where they can escape from human and other visual and acoustic contact throughout much of their pregnancy. One can provide the public the educational opportunity of watching progress in pregnancy and birth of cubs by using infrared cameras and closed circuit television. This can also allow the husbandry and veterinary staff to observe the animals while eliminating or greatly reducing intrusion on the pregnant female.

### Protecting Animals and Delicate Electronics from Electrical Shock

Our work has greatly benefited from the availability of inexpensive opto-isolators. These may be purchased from electronic distributors (or often much less expensively from surplus outlets). The general principal is that devices can be interfaced by means of light emitting diodes and photocells, thus eliminating the need for direct electrical connection between high and low voltage apparatus. A good enrichment planner can design equipment that isolates both animals and low power electronic devices from dangerous electrical power levels by using opto-isolaters.

## New Designs for Animal-Powered Devices

The following devices were designed for use in enrichment efforts in animal care facilities which are in biomedical research settings. In most cases they could serve well in enriching the life of any captive animal(s) by empowering them to control treat delivery while entertaining themselves and getting a little exercise. Devices that use batteries or other power sources require scheduled maintenance, complicate installation, and may restrict choice of placement of the enrichment apparatus. Using animal powered devices eliminates these concerns.

On one of the many occasions when I thought I was ready to wrap this book up for publication, my friend and frequent current apparatus collaborator Rick Jarvis came up with a concept for a vertically rotating food delivery wheel. As is often the case, his good idea kept me awake most of the night, designing in my head ways to make his basic notion into safe, reliable, affordable means to enrich the lives of some animals.

We worked together on the design, and Rick manufactured the device that is shown in figures 20-9 through 20-13. This device is used for special kinds of work, that my university coworkers and I have undertaken. Our goals are to improve the well-being of primates that are in need of exercise, and to help to bolster cooperative behavior between social primates which we hope to move from single housing to larger social housing. Figures 20-12 shows the apparatus mounted on a large cage specifically designed to allow more exercise than standard cages. It is also large enough to house 2 social primates, and allow them to move around easily. Figure 20-13 shows a macaque who has learned to get treats when he wants by pulling successively on two widely spaced toys that descend into his quarters. This encourages exercise and it also entertains the macaque. Like most mammals, he also clearly enjoys seeing what he can control.

Figure 20-9. Front View of Ratchet Driven Rotary Enrichment Device

The recipe below is one for a newer version of the apparatus. I designed this version to have many advantages when compared with the prototype seen in the illustrations. Among the advantages: 1) It can be easily attached to the outside of any structure or cage housing captive animals; 2) It can be easily changed for applications where the resident animal can use the ratchet handle directly to operate the devices, to applications such as the one seen in figures 20-12 and 20-13 where the device is operated by pulling successively on two separate cables to advance the ratchet; 3) *And*, it is much easier build.

Please carefully read *all* of the steps in the recipe before purchasing parts. Also, please keep in mind that this latest design uses a slightly different arrangement of components. The parts needed for construction are fewer than the number used to make the original version of the ratchet driven device. Consequently, while I believe the illustrations will be of use in understanding the basics of how the mechanism works, they will not match precisely with the recipe instructions.

If you plan to wash the enrichment device with high temperature liquids you will have to use high temperature rated plastic and high quality stainless steel machine screws, set screws and nuts.

**Recipe 1. A Simple Inexpensive Ratchet-Based Feeder**

The word(s) following colons in the following list indicate the brief connotative name that will be used for parts in the recipe.

Items Needed for Manufacture

Quarter inch thick transparent rigid plastic material to make the following parts for the enclosure: "top", "bottom", "back", "front", "left side", "right side", "partition"
A sufficient length of 1½" inside diameter schedule 40 plastic water pipe to manufacture: "silo", "cup"
One end cap for 1½" schedule 40 plastic water pipe: "silo lid"
One six inch schedule 40 PVC pipe end cap with a depth of at least 2½" and a flat end: "wheel"
A ¾" length of thick walled ¾" PVC pipe: "pipe collar"
One high quality ¼" socket drive ratchet with a handle that is at least 7" long: "ratchet"
One socket that fits a ¼" ratchet drive, and has an outset diameter that will fit tightly into a ¾" diameter hole: "axle socket"
One short ¼" ratchet extension bit that has a round shaft: "axle"
One ½" x 1" x 6" plastic block: "ratchet excursion limiter"
One 3" x 2 ½" x ½" plastic block": "silo retainer"
One 3" x 2 ½" x 2" machinable plastic block of plastic material that will provide a good bearing surface: "cup bottom seal"
One 2" x 8" x 1" machinable plastic block: "axle bearing".
Stainless steel machine screws, set screws and nuts

1. Cut a one inch length of the PVC pipe to make the cup, and an eleven inch length to make the silo.

2. Clamp the wheel in a position so that you can drill a close-fitting hole into which you can install the cup. The hole must be perpendicular to the rim and located approximately ¼" in from the open side of the wheel (figures 20-9 and 20-10).

Check to make certain that the cup fits tightly in the hole by pushing the cup into the rim. Take the cup back out and daub PVC cement in the hole in the rim to lock the cup in place. Reinsert the cup *while being certain that the top of cup matches perfectly with the inside surface of the rim, and the rest of the cup protrudes outside the wheel.* Machine or file the top edge of the cup to match the curvature of the inside of the wheel rim.

3. Note that in figures 20-9 and 20-10 the food silo is sloped at an angle so that it can cover the top of the cup, and just clear the outside edge of the wheel rim. Machine the bottom of the silo to the slope and curvature required to fit flush with the top of the cup. Hold the silo in place over the hole, and carefully measure and record the angle at which the silo slants away from vertical.

4. Make a threaded 1/8" diameter hole in one side of the pipe collar, ¼" inch from one end. Screw a 1/8" diameter set screw into the hole. Lay the wheel down on its open side. In the exact center of the side that is now facing up, drill a hole that will tightly fit the pipe collar. Use PVC cement to glue the collar in place, with one end of the collar extending out enough to allow access to the set screw.

Push the axle socket into the pipe collar until the end of the socket that will attach to the axle is flush with the edge of the collar. Tighten the axle socket in place with the set screw.

5. Precise machining of curved surfaces on the cup bottom which protrudes from the wheel, and on a 3" wide surface of the cup bottom seal, is necessary to prevent spillage from the cup as it passes over the cup bottom seal.

Make a template for machining the cup bottom seal, and the cup to the curvature necessary to prevent leakage from the cup bottom as the cup rotates over the cup bottom seal. Hold the cup bottom seal with a 3" x 2" surface facing up, and lay it on a flat surface. Use the template to mark the curvature required, and machine the cup and the cup bottom seal. Check to ensure that when you rotate the wheel, the bottom of the cup does not leak when it is sealed by the cup bottom seal.

6. In this step the enclosure sides and partition will be cut from the ¼"transparent plastic, and assembly of the parts will begin.

Cut the front and back each 12" high x 10 ½" wide.

Attach one end of the axle onto the ratchet, and the other end into the axle socket that is fastened to the wheel. Place the wheel in an upright position, and insert the silo over the top of the cup as described in step 2 above. While holding the silo at the necessary slant to cover the top of the cup, measure the horizontal distance from the outermost edge of the top of the silo to the outermost part of the ratchet. This distance is referred to as "D" in the rest of this paragraph.

Cut the left and right sides each 12" high x D+1" long. Cut the top and bottom each 10" wide x D+1" long.

Use stainless steel screws to attach the sides to the edges of the top and bottom. This will produce outside dimensions of 10 ½" wide x 12"high for *later attachment* of the front and back of the enclosure. Do not glue the enclosure together, since you will need to occasionally remove a wall or two of the enclosure in order to clean or adjust interior parts of the apparatus.

Make the partition by cutting a rectangular piece 11 ½" high x 10" wide, so that it will fit the inside dimensions of the enclosure when it is installed in parallel to the front and the back. Directly in the center of the partition, cut a 2" wide rectangular opening extending from the top of the partition to 1 ½" above the bottom. The partition will serve to keep the wheel in proper alignment to prevent stress on the silo when the ratchet is operated.

7. Mark the center of a 3" x ½"side of the silo retainer with a line that is slanted at the angle necessary to properly align the silo in the apparatus. Drill a hole to insert the silo centered in a 3" x 2 ½" side of the silo retainer, and slanted at the angle that you have marked on the ½" side.

Place the silo retainer on top of the enclosure, centered from side to side, and 3/8" from the front edge. Ensure that the hole in the silo retainer slants toward the inside of the enclosure.

Use machine screws to attach the silo retainer to the top of the enclosure. Use the same size drill bit as you did for the silo retainer, and insert in the hole in the silo retainer so that you can drill the top of the enclosure to allow inserting the silo through the retainer and into the enclosure. This will allow you to easily adjust the silo to rest over the cup when all parts are in place, and then lock it in position with the set screw in the silo retainer.

8. Place the cup bottom seal a few inches inside the enclosure and centered from side to side. Insert the wheel into the enclosure while holding it in a position with the cup at the bottom and the open side of the wheel facing the front of the enclosure. Rest the cup on the cup bottom seal. Insert the silo through the silo retainer. Slide the silo into the enclosure and inside the wheel.

Adjust the position of the silo and the cup bottom seal until you position the silo so that it will exactly cover the entire top of the cup. Insert the enclosure partition in the front of the enclosure, and slide it in until it rests on the open side of the wheel. Check to make certain that the silo is still aligned perfectly to cover the top of the cup, and attach the partition solidly in place with machine screws.

Carefully move the cup bottom seal until it is against the partition, and use machine screw to attach it to the partition.

9. Attach the axle to the wheel. Insert the axle bearing into the back of the enclosure, and hold it alongside the axle with the 1" side of the axle bearing touching the axle. Precisely mark the height at which to drill the hole in the axle bearing.

The bearing hole must allow the axle to rotate smoothly, and be tight enough to prevent the axle from wobbling. Place the axle bearing laying on a level surface, and drill the hole centered from side to side at the dimension you marked in step 9.

Figure 20-10. Side View of Ratchet Driven Rotary Enrichment Device

10. Unsnap the axle from the axle socket, and insert the axle in the axle bearing. Snap the axle back into the axle socket. Slide the axle bearing toward the wheel until it as close as possible to the wheel, while still allowing the wheel to rotate with minimum friction. Use machine screws to attach the axle bearing to the bottom of the enclosure.

11. In this step you will make a 5" high x ¾" wide vertical slot in the left side of the enclosure to allow the ratchet handle to protrude from the left side. Snap the ratchet onto the axle, and the axle to the wheel. Carefully measure the distance from the end of the ratchet to the back of the enclosure. Mark the left side of the enclosure so that the ratchet handle will protrude from the middle of the slot, and the slot is centered from top to bottom of the enclosure. Produce the slot with a router or carefully cut it out with a saw.

12. Measure the thickness of the ratchet handle. Mark the ratchet excursion limiter, so that you can machine a 5" vertical slot just slightly wider than the thickness of the ratchet handle, and centered both vertically and horizontally in the 1"x 6" dimension.

13. To allow adjustment of how many times the animal must operate the ratchet in order to receive each treat, the ratchet excursion limiter must slide up and down while the handle is centered in both slots. One means to accomplish this is to make 2 vertical slots as close as possible to the sides of the limiter and tap holes in the side of the enclosure for 6-32 screws.

14. Snap the ratchet off the axle, and pass the ratchet handle from the inside of the enclosure through the slots in the side and the excursion limiter, and then attach it to the axle again. Place a few treats in the silo, and test the operation of the device. After making any adjustments that may be necessary, screw on the front and back of the enclosure, and you should have the finished product.

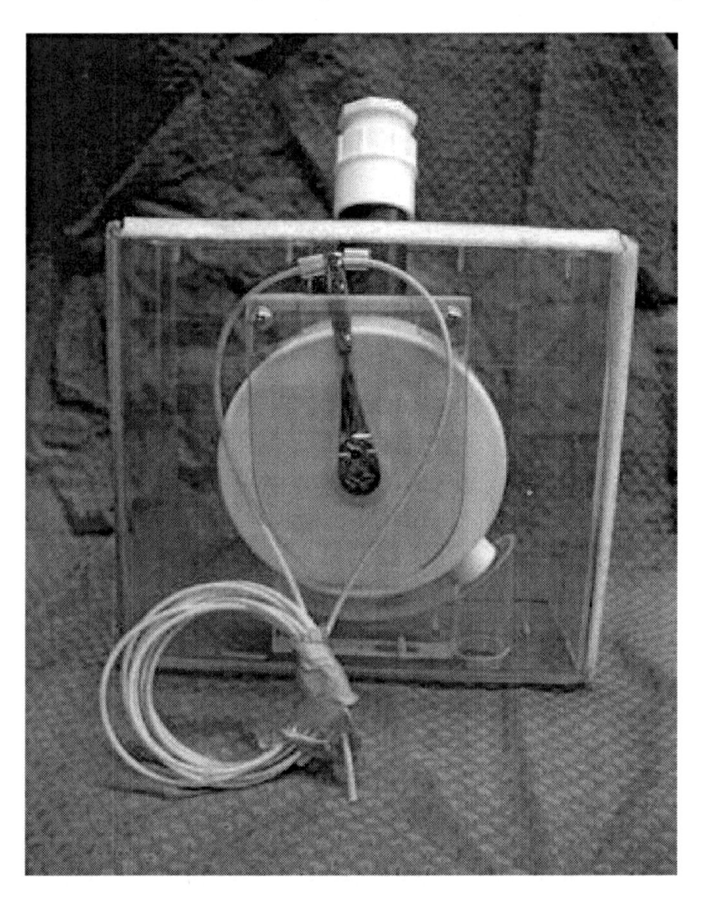

Figure 20-11. Ratchet Based Device with Cables That Can Span Six Feet

I like this new design because it only takes a little imagination to think about ways to adapt it to a variety of enrichment uses. I feel sure that inventive readers will quickly think of many, but as a teaser I offer a couple that immediately came to mind as I was writing this section. Many animals such as guinea pigs, mice and rats, whether they are pets or animals employed in research, enjoy running in exercise wheels and derive health benefits from the activity. It would be an easy matter to make removable spokes that could protrude from the perimeter of the wheel. These spokes could be positioned to engage the ratchet. Then you could add or remove spokes to determine how many wheel revolutions were required to gain a portion of treats, and remove spokes at random to produce less predictable delivery of treats.

Figure 20-12. Ratchet Device with Cables Used to Span Large Enrichment Enclosure

Figure 20-13. Macaque Pulls on Toy to Advance Ratchet Based Treat Feeding Device

A colleague at the University of California, Davis was interested in potential applications of the device for his research with birds. I sketched a modification with threaded holes for stainless steel screws concentrically arranged close to the perimeter of the end cap. The heads of the screws would be used to activate a push type toggle switch to turn sounds off and on, which is one of the things this friend would like to do to empower birds involved in his research. By adding or removing some of the stainless screws, variety could be added in myriad ways for the birds, such as activating different kinds of sounds for a relatively unpredictable number of pulls with their beaks. For this purpose the device could also be more simply constructed without the food treat delivery parts, but I liked the idea of being able to see what the birds preferred by trying different paradigms. For example you could try associating some sounds with treat delivery, or do something more potentially interesting for the birds, such as having a sequence in which rotations of the wheel activated increasingly exciting sounds until food was delivered.

**A Modified Version of the Rooting Device Described in Chapter 12**

Rick Jarvis and I have been working on a compact version of the rooting device that I had first used as a treat feeder for pigs more than twenty-five years ago (see chapter 12). One addition for some uses would be to increase the maximum number of rotations of the rooting wheel necessary to earn treats. This might be desirable for use with animals that have become so proficient at earning treats that they quickly expend the reservoir of treats. By making the ratio of rooter rotations to treat delivery greater, it is possible to encourage rooting activity for a more extensive part of the day. Figure 20-14 is Rick's conceptual illustration for how this could be accomplished.

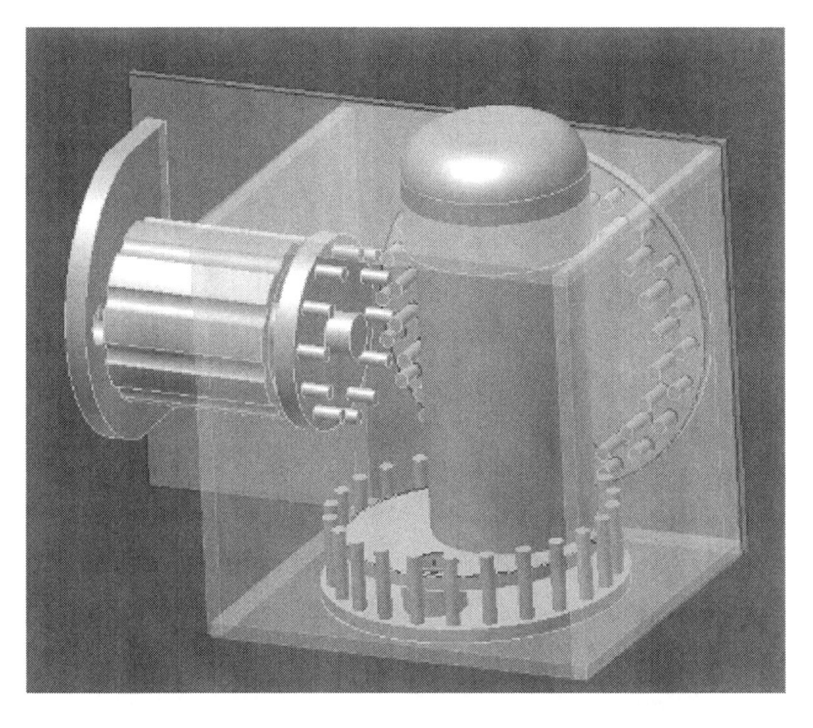

Figure 20-14. Conceptual Drawing of Apparatus that Allows Enrichment Workers to Select Larger Ratios of Reinforcement to Encourage Activity

For most uses of the apparatus in our enrichment work, we have found that the simpler original device described in chapter 12 serves well and is easier to make reliable and sturdy. Our major focus is on producing rugged, reliable, trouble-free, apparatus that provides opportunities that captive animals find attractive in controlling parts of their environment.

**One Last Idea and Simple Recipe for Encouraging Cooperation between Primates**

Chronologically this recipe is the last thing I am writing before wrapping up the book, and sending it off for publication. It would not have been written except that this past spring my efficient and technically skilled friend Ryckje, who was helping me with final formatting and assembly of the manuscript, fell down the steps and broke her arm and wrist. I hope it will not bore you if I share the genesis of this brand new invention for which we have great hope in our work with laboratory primates. At the very least it should be of interest in illustrating what sort of things transpire in a weird mind.

While waiting for Ryckje to heal for a few months, I was at my lab one day, and while on a break and looking at manuscripts sent to me for review by my past students, I began to think of the days when they were in my lab or in the field with me. Most were very easy to work with cooperatively in research, but a small number were essentially loners because they were so highly competitive and demanding of others. I mused on the fact that we were not the only animals for which there was this kind of variation in demeanor. In all of the mammalian species that I have closely studied in the field or lab, there are apparent kinds of behavior uniquely associated with individuals. Whether because of genetic makeup, events in their past, or a combination of these factors, some individuals just have a hell of a time cooperating or mixing socially with their own species.

That led me to think about a lab experiment that so intrigued me over the years that I had assigned some of my slaves (students in laboratories required for my courses) to rerun the experiment frequently. It was a simple test of how conspecifics of a variety of mammalian and avian species would act in what is best metaphorically described as a "you scratch my back and I'll scratch yours" situation. The experiment involved separating two animals within an enclosure by means of a transparent divider, and providing each a method to feed the other animal. Animal A could only be fed by animal B *after* A had operated some kind of device to feed B, and vice versa. The results were always of interest to the students.

In one of the very first labs studying the development of cooperative behavior, the students chose to work with rats and pigeons. Many of them reported with pride how successful 'their animals" had been in cooperating. The two students working with the only ultimately unsuccessful pair asked me to share with them in bemoaning the fate of rat "A" who did his damndest to cooperate with the other. When A fed B, B would chomp the food down. But when B pressed a bar and A was delivered a treat, B would literally splay himself against the divider in apparent frustration at seeing the other rat consume his earnings. Regardless of their patience and repeated efforts to demonstrate by pushing the bar for B to initiate cooperation, after six hour long sessions B never learned to work in reciprocity. (Any conscientious teacher will understand the sadness expressed by the rats' mentors!)

A few days after reminiscing about these events, I was talking to a colleague at UCSF about the difficulties in enriching the lives of captive primates by allowing them to live in pairs or groups. The only university laboratory animal care facility that reported unbridled success in this area was one in which primates were essentially treated as pets. Lab personnel regularly came into direct physical contact with non-human primates while wearing no protective gear. Because of potentially lethal diseases that can be transmitted between non-human and human primates, today this kind of unprotected contact is prohibited in most animal care facilities.

We have found that using techniques like inserting temporary dividers, and closely observing primates as opportunities for physical contact are gradually increased to be of benefit. But this is very time consuming and does not always work. I suggested that it would be really nice if there was a reliable inexpensive way to encourage cooperation and more adequately assess compatibility *before* allowing primates direct contact with each other. I mentioned the frequently conducted lab experiment described, and my wish that we could employ the same general procedure without expensive and cumbersome equipment. Inexpensive reliable equipment which facilitated primate socialization would be used in many primate care facilities. I surprised myself by suddenly thinking of a concept for such a device.

During the following month I made rough drawings, scale models, and templates to test various ideas before coming up with a design and writing specs for the prototype. The instructions for building the prototype are reproduced below. Three are currently being tried in animal care facilities.

Although it requires a good craftsperson who can work carefully to these specifications to make it reliable, the apparatus is not complex. It can be made of inexpensive materials, is light weight, and is adaptable for many kinds of captive animal quarters. I specified high temperature transparent plastic and high quality stainless steel screws (SS316) for use in care facilities that use high temperature cage washers to sanitize all equipment. This increases the cost of materials considerably (total cost = 250 dollars total from a reliable distributor of new products). Because I know that many folks interested in effective methods of enrichment have tight budgets and are willing to do all apparatus cleaning manually, I did some quick research online. By purchasing remnants and overstock plastic and standard quality stainless screws, you can obtain the supplies to build a long lasting device for about fifty dollars.

Figure 20-15 shows the general concept of the device without showing the inner parts that are visible through the transparent plastic elements. When the plunger at the left end is pushed all the way in, food is delivered on the right end and treats are loaded for delivery at the left end No further responses can be made until the plunger at right end is pushed in reversing the action, delivering treats at the left end and loading for next delivery at right end. The length of the device can be changed for specific environments, as can the height of the silos, so they are shown in figure 20-15 in cutaway format.

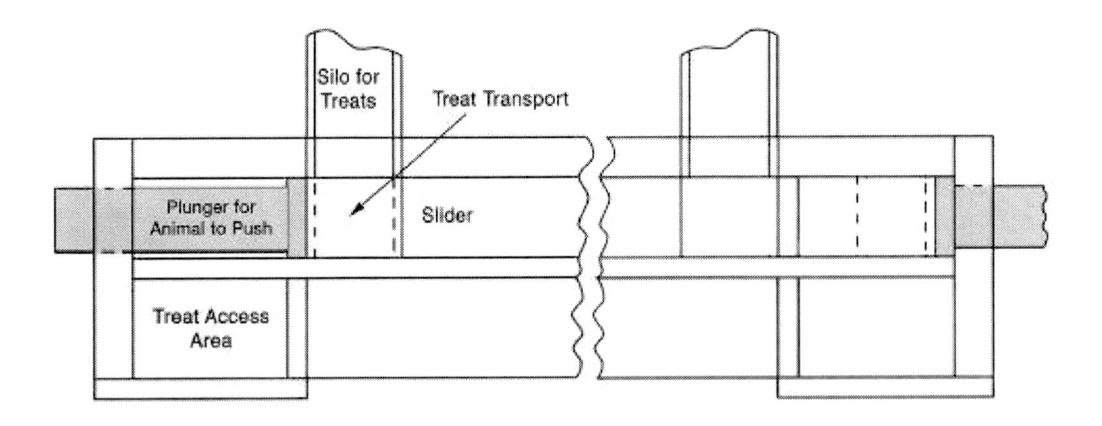


Figure 20-15 Side View of Apparatus for Encouraging and Testing Cooperative Behavior.

**Simplified Instructions for Building Co-op Feeding Device**

Figure 20-16 shows the assembled parts with labels for use in following these instructions. Parts shown in parenthesis are corresponding parts for the unshown side and end of the apparatus.

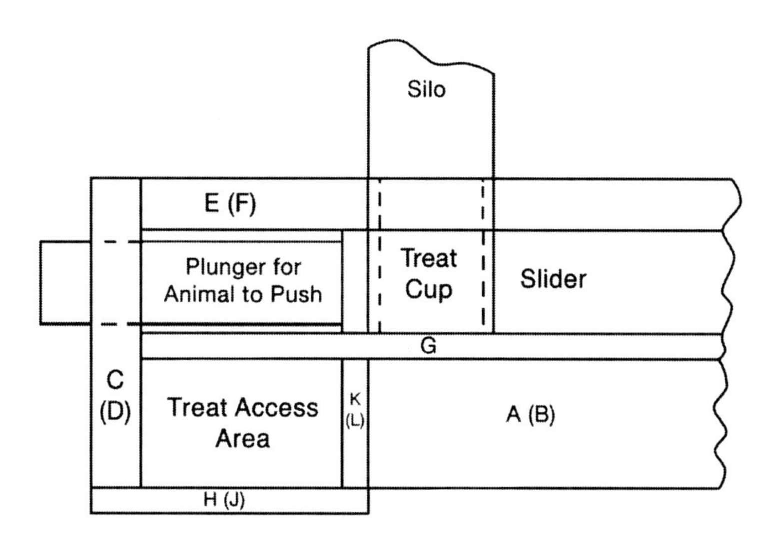


Figure 20-16. Parts Identification for Device Assembly.

There are several places where it is critically important that measurements be as precise as possible in order to produce a device that will work without jamming and require little maintenance or repair. These places are marked with asterisks.

1. Cut two 3" strips from ¼ inch thick x 24 inch wide plastic (parts A & B).

2. Refer to figure 20-17 to cut part A to proper shape by removing the two small rectangular areas as shown. Part A goes on the side where the animals accesses treats delivered by their cooperating partner.

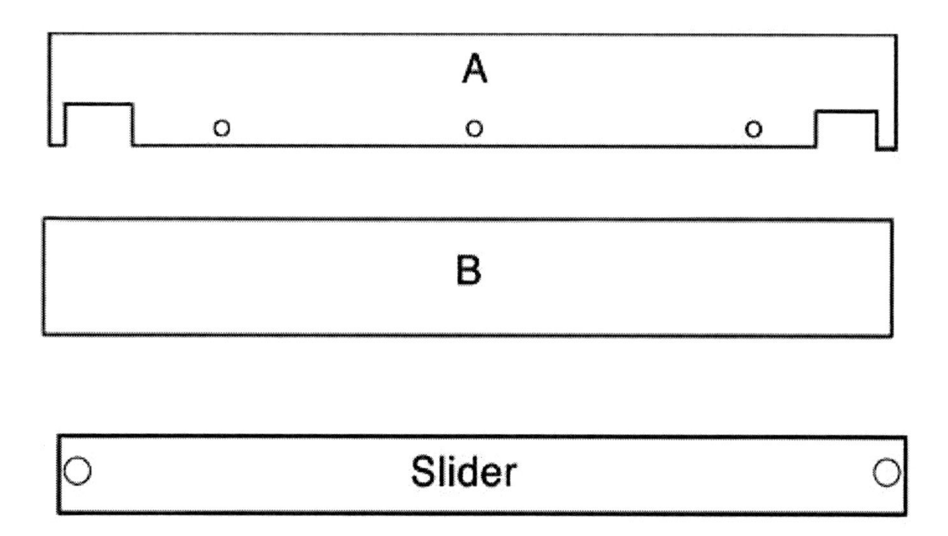


Figure 20-17. Parts A & B: side views; Slider: top view

3. Drill three 3/16" inch holes with centers 5/8 inch up from the bottom of part A, one at 5 inches from each end and one at the middle. These holes are to use J hooks to easily attach the apparatus to cages.

4. Cut two 3 inch lengths of ¾" rod. Drill and thread a hole in the center of one end of each rod so that stainless screws will fit tightly when screwed in.

5. Make two rod retainers 1/4" thick x 3/4" x 1". Make a countersunk hole in the center of each retainer, just large enough for the stainless steel machine screws. Firmly screw one retainer to the end of each rod. *The screw head must *not* protrude from the retainer after it is screwed all the way in. The part you have assembled in this step is labeled "Plunger" in the diagrams.

6. Cut two 3 inch lengths of ½" x 2" nylon block (parts C & D). *In each of these two parts, drill a hole centered in the 2" dimension and one inch from the end and just large enough to allow the ¾ inch rod to move easily. Insert one plunger in part C and one plunger in part D.

7. Cut a 12 inch length of ½" x 2" nylon block in half to form parts E & F.

8. Refer to figure 2, and prepare to assemble all pieces made so far using high grade stainless steel 4/40 machine screws to attach A, B, C, D, E, and F. Mark appropriate places for screws spaced no more than two inches apart. Use a 4/40 tap to thread the holes to snuggly fit the machine screws using.

9. Screw parts together temporarily.

10. Make the part labeled "Slider" by cutting a 1" x 2" x 24" block to a length of 20 ½". See if the Slider will fit snuggly below parts E &F in the assembly. *Do not force it in if it does not fit.* If it does not easily fit snuggly into the assembly, loosen screws as necessary to allow it to snuggly fit. There should be some resistance in movement as you test by pushing the plungers to move the slider back and forth between ends of the assembly.

11. *Drill one inch holes in the slider centered on the 2 inch surface and centered ¾ of an inch from each end. These two holes will form the "food cups" which are within ¼ inch of each end of the slider.

12. Measure the rectangular opening formed at the bottom of the enclosure. It should be approximately 23 inches long and barely wider than the slider if assembly has been properly accomplished. Cut a piece of ¼ inch thick plastic to fit the opening (part G). G will be the surface upon which the slider moves.

13. Remove the slider and part G. Carefully use a carpenters' square to see that one end of the slider is lined up with one end of G and use the hole in the slider to scribe a circle in G. Repeat this step for the other end. *Drill one inch holes in G where you have scribed the circles.

14. Assemble all of the parts manufactured thus far and check for accuracy and reliability of operation of the slider. When pushing on the plunger, there should be appropriate resistance from friction of parts so that the primate can easily push the slider in, but not move it so freely that it bounces off the opposite end. If necessary, adjust tightness of screws to accomplish the required resistance. Double check to make certain that the hole in the slider perfectly lines up with the hole in part G when you push the

plunger all the way in. This alignment is critical for smooth operation of the device and to ensure that treats from the "cup" will empty into the access area so that monkeys can gather them.

15. Cut two 12" lengths of 1" transparent plastic pipe to make the parts labeled as "silos".

16. Invert the apparatus and push the slider all the way to one end and make sure the hole in G lines up with the "cup" hole at that end. *Remove G and while keeping the slider firmly in location, use the cup hole at the *other end* of the slider to scribe a circle on the part above the slider (E or F). Move the slider all the way to the other end and hold it in place while scribing a circle in similar fashion on the part now in line with the cup hole at the other end.

17. Remove E and F and center the silos on the scribed circles. Keep each silo firmly in this position and use the silos to carefully mark for drilling holes through E and F. Make sure that the holes you make will snuggly fit the silos. Use an appropriate adhesive to glue the silos in place while being certain that they *do not protrude* beyond the bottom of the holes you have drilled for them. Do not let any of the adhesive fall onto the surface of the slider.

18. Cut two pieces of ¼ inch thick clear plastic approximately 2 ½ inches square to form parts H and J. Screw these parts to the bottom of the apparatus on each end as shown in figure 20-15 to form the bottoms of the access areas.

19. Cut 2 pieces of ¼ inch thick clear plastic, each measuring 1 ¼ x 2 inches to form parts K and L. Screw these parts in place as shown in figure 20-16. The access areas will now be open on the side of the apparatus that faces the animal, and the remainder of the access area will be enclosed to prevent spillage.

20. Use any small non-sticky food or other objects measuring less than 7/8 inch to fill the silos, and test operation of the device.

The reason that I have added this recipe and further delayed publication is because I am excited by its potential for use in enriching the lives of so many captive animals. With just a little ingenuity, one can easily alter the design for those kinds of animals that cannot easily push a plunger. It is such a simple and inexpensive device that facilities with low enrichment budgets will still be able to afford to produce some. And, the same apparatus can employed in a wide variety of captive situations.

My immediate plan for the prototypes involves assessing the utility of the apparatus in our efforts to improve the lives of primates housed in three different arrangements in research facilities. For individually caged primates, in reasonably large cages, we will make the device as long as possible. Once the resident animals learn that they can earn treats by moving from one end of the cage to the other and pressing the plungers at each end, we will evaluate carefully whether it produces healthful activity. If only one silo is loaded, the animals will only get a treat for the complete cycle between ends. If both silos are loaded, treats will be deposited at the opposite end only if the primate presses the plunger after consuming the nearby treat. We can then ascertain which method ultimately encourages movement to improve the primate's vigor. We captive primates always need some kind of encouragement and payoff to get us to exercise!

In the second circumstance, primates that have been housed individually for long periods will be placed in adjacent home quarters. The device will span the two cages and be used to encourage cooperative behavior. We will evaluate whether the use of the apparatus improves the chances of success in pairing primates when compared other pairing procedures. We may also be able to identify in advance which animals are simply unwilling to learn to cooperate and share resources.

Thirdly we will introduce the device, in environments where animals are already group-housed. Then we will evaluate effects of providing animals the opportunity to cooperate in feeding each other treats whenever they wish. Of course, some primates who do not welcome the chance to work cooperatively may choose to do the work at both ends of the device on their own...as was the case with a few of my old grad students.

For primates unwilling to learn to use the device efficiently by cooperating with their peers, enrichment workers can teach them the use of the device by being their partners in learning. If you want to make this teaching opportunity really popular with your enrichment staff, load the silo that fills the cup serving rewards to the teacher with coins!

**Closing thoughts about apparatus**

There are many new devices that have arisen as time has passed in creation of this book. Because of the astonishing rate of development in microelectronics, efficient tiny electronic devices can replace many of the much larger and less efficient components used in the invention of many of the devices described in this book. Every day I see new information about less expensive, more compact, more flexible and more powerful devices that can be used in enriching animal lives. For example, today I am working on a device that employs a commonly marketed electronic device with a touch screen, and modifying it for use by non-human animals. The animals will be empowered to select between wide varieties of different kinds of audio-visual options. Sensitive caregivers will enjoy selecting materials and observing which ones are most enjoyed by animals in captivity.

It is my sincere hope that you will find this chapter helpful in your enrichment efforts. However, it is only your imagination and the skill of your collaborators that limit the range of devices that can be conceived for use in empowering captive animals. I hope that you will share your own inventions with all who work to enhance the lives of animals.

# 21

# GRAPHICS FOR ENRICHMENT

Readers of this book may well be the sort of folks who voraciously consume every new bit of information they find in zoo graphics. Graphic designers award each other with accolades for truly inventive exhibit graphics which should excite all viewers. But the sad truth remains that the majority of zoo visitors spend little, if any, time reading graphics.

However, if you have created active enrichment opportunities which energize animals in captive facilities, a fringe benefit is that multiplicatively more visitors will search out and read graphics. They wish to be able to tell others why the serval is suddenly leaping in the air, why the previously sleepy leopard is running up and down the tree limb, and more about the natural skills of the animals they are observing. If you have arranged a means for the public to somehow participate in the enrichment activity, they will doubtless read to learn what they can do to interact. In this chapter I will describe several examples of graphics that we have successfully used and suggest a few others to stimulate your efforts to accompany enrichment for animals with public education about the species' abilities.

**Wordless Graphics**

Sometimes the very best graphics are those that have the fewest words. For example, I remember a great often-read graphic that Bill McCabe made for the Portland Zoo. Rather than a lengthy treatise on recessive characteristics, Bill's attractive pictures illustrated with large arrows the genetic contribution from parents that would lead to the Black Panther that they could observe on display.

I learned from working with Bill that simple wordless information often worked well.

Figure 21-1. A "Bird in the Bush" Showed Visitors Why the Leopard Was Hunting.

Much later, when planning graphics for the acoustic based enrichment used to enhance the life of a leopard in the San Francisco Zoo (Chapter 3), I included a very simple wooden bird (figure 21-1). The "bird" was located in the area where chicken morsels were delivered to the leopard for successful pursuits. This was sufficient clue to allow visitors to tell others to "look where the bird is in the bushes" thus effectively reducing the need for wordy graphics.

**Illustrations Associated with Special Visitor Opportunities**

Many zoos have benefited from Merv Larson's brilliant exhibit designs, which were incorporated in the early development of the Arizona-Sonora Desert Museum. Some of these exhibits used carefully planned lighting to allow the visitors to see aspects of the private lives of animals such as behavior in dens. For example, small windows were installed to allow visitor to see into a den, and careful design minimized intrusion on the animals' privacy. Visitors observed from dark areas where they were virtually invisible to animals occupying the den.

The graphics in Figure 21-2 show inventive use of the same technique in the High Desert Museum in Oregon, where Don Kerr championed the idea that animals should be maintained in as naturalistic areas as possible. During my consulting trips to help Don develop naturalistic enrichment plans for otters, I observed how much special attention was drawn to this opportunity to learn about the private lives of these animals.

Figure 21-2. The "Private Lives" of Otters

Careful observation yielded clear evidence that visitors almost always read graphics at exhibits such as the Asian Otter exhibit at Marine World, where the graphics showed them how they could participate in the enrichment activity (Figure 21-3). Imbedding the illuminable buttons for public participation within the graphic panel itself allowed the visitor to participate in the active opportunities we had provided for otters, without stepping away to read some lengthy instruction.

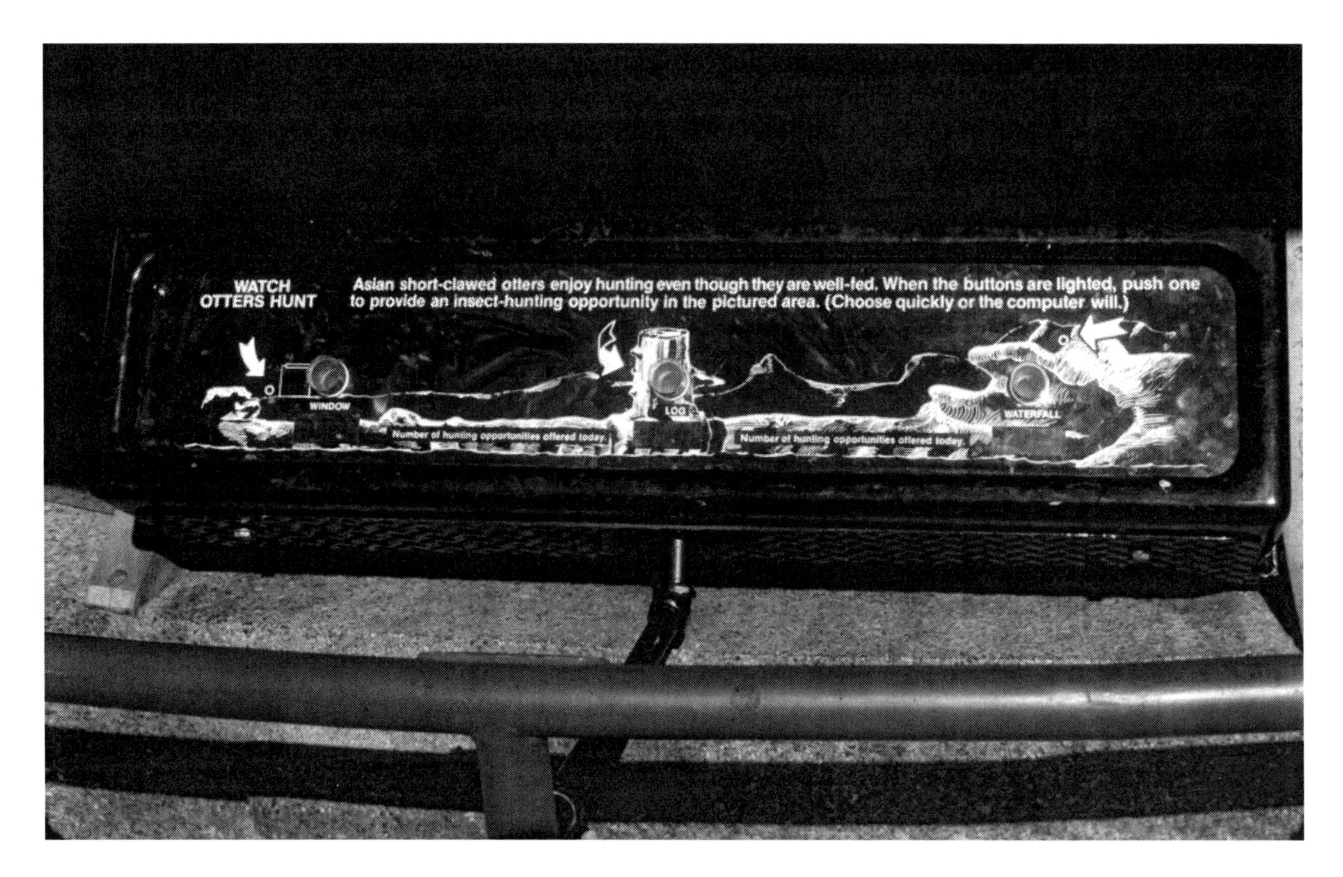


Figure 21-3. Otter Cricket Hunt. Meaningful active participation with otter enrichment requires learning a little bit about what is happening and why.

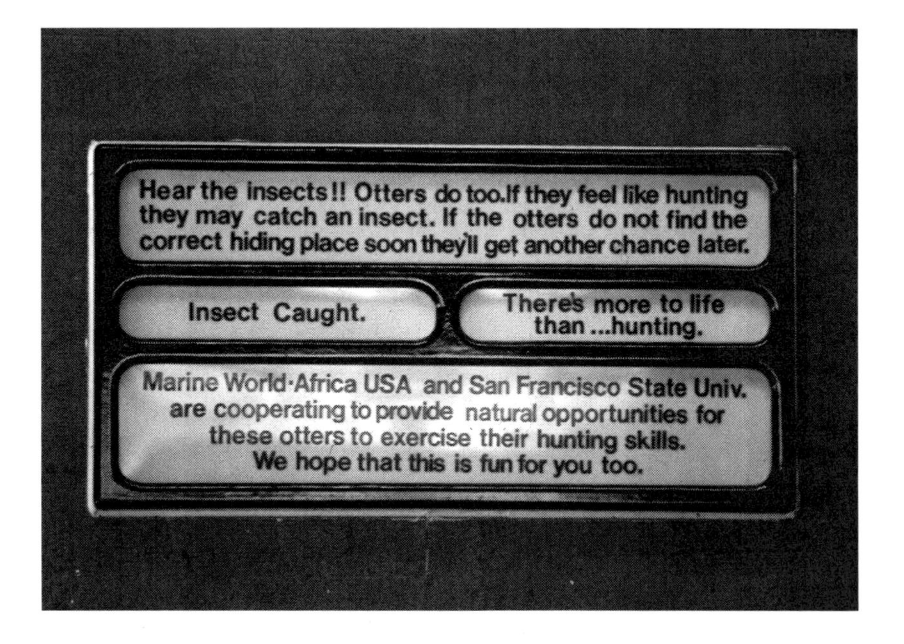


Figure 21-4. People attracted by the sounds of prey and sight of otters scurrying in the hunt would gather at the exhibit and study the graphic to see how the hunt was progressing.

Of course, only one visitor at a time could actively participate in the enrichment effort, so for exhibits such as this we used dynamic graphics to let others know how things were progressing. A good example of this appears in Figure 21-4. This transluminable panel successively lighted in appropriate sections as the animals' hunt progressed.

I will conclude this brief chapter with two examples from our early efforts which I personally feel very differently about. The first procedure I would never again advocate for reasons I will describe below. The second was the finest effort that Ron Fial and I accomplished in our collaborative efforts to provide naturalistic opportunities for animals, and foster education about species-typical behavior.

**Graphics for Unnatural Games**

Our band-aid effort to provide a solution for a serious problem for mandrills (*Mandrillus sphinx*) housed in a mausoleum type cage in the Portland Zoo in the 1970s has been described in detail in a number of my articles and book chapters (e.g., Markowitz 1982.) The results of this effort were featured in national television shows, newspaper features, and magazine articles (figure 21.5). Despite the fact that this brought a surprising amount of positive response and publicity to the Portland Zoo, it always left me with a queasy feeling. It did little to teach the public about the need to conserve these wonderful animals, and it did nothing to exhibit the special abilities that Mandrills would exhibit in natural surroundings. What it *did* accomplish was to improve the well-being of the cage residents (Yanofsky and Markowitz, 1978) and illustrate to visitors that we are not the only primates capable of competing effectively in games that required attention and skill.

Figure 21-5. Visitor versus Mandrill in a Speed Game Contest

One graphic panel showed the progress of each participant (figure 21.6). It also indicated whether the Mandrill was playing against a human who had chosen to participate within a predetermined time, or if the Mandrill was playing against an electronic device.

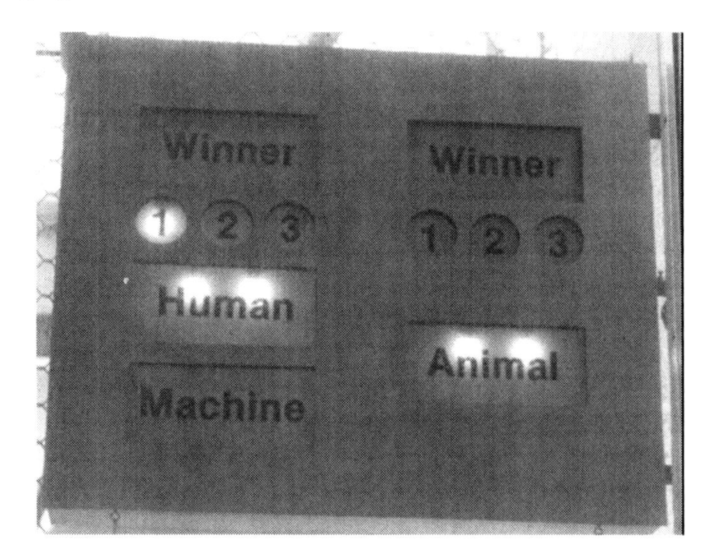


Figure 21-6. Transluminated graphics show progress of competition between mandrill and human or computer in a speed game

Another graphic encouraged participation (figure 21-7), and also explained that if no one chose to donate to expanding these efforts by inserting a coin to compete, the mandrill would not be deprived of opportunities to compete because the computer would shortly compete with him. We used coin-operated apparatus on a few exhibits to raise money in order to expand our enrichment efforts to additional groups of primates. This approach brought in thousands of dollars that allowed us to progress to further enrichment work.

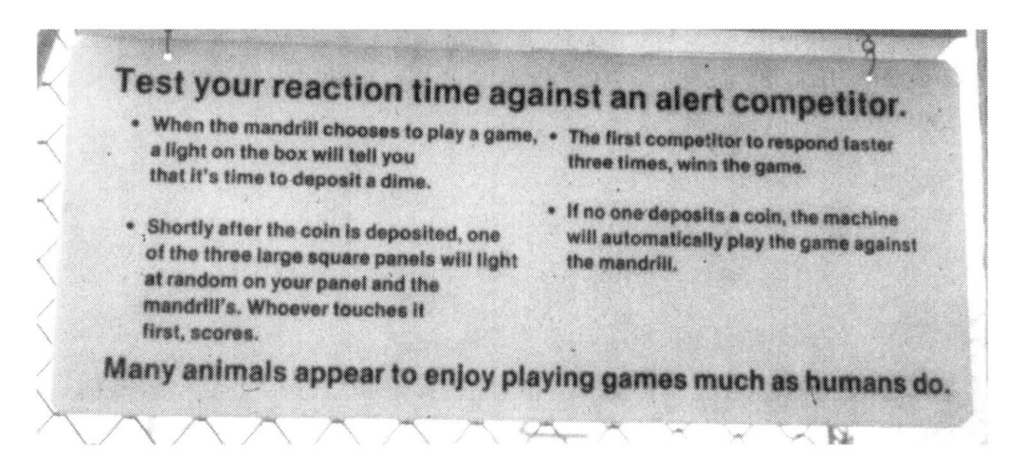


Figure 21-7 Graphic explaining participation in contest by humans or between computer and mandrill

I was increasingly saddened by the fact that the public seemed to largely ignore the parts of instructional graphics that explained that animals were not machines, and they might choose not to respond. It pleased me greatly that when we provided other opportunities for public participation in enriching the lives of captive animals in places such as the Honolulu Zoo, Marine World Africa USA, and the Panaewa Rainforest Zoo in Hilo, we were able to abandon the coin operated approach to providing sufficient funding. It may have been a minority of the visitors who had a 'pinball game mentality' about depositing coins to engage in some activity such as competing against the Mandrill. But I felt that this approach to gaining monetary support for enrichment efforts was in many ways demeaning to the resident animals, and I would never be involved in this type of activity in the future.

**Graphics to Accompany and Encourage Participation in More Naturalistic Efforts**

Even though it only had a brief life because of financial difficulties that plagued the Panaewa Zoo, I have always felt that our best effort with a limited budget was seen in the work that we did there (Markowitz, 1982). One of the things that I especially liked was the dynamic graphics that Ron Fial and I devised to ensure an educational as well as an entertaining opportunity for visitors to participate in various stages of this enrichment work.

In figures 21-8 Ron is seen installing and testing components of the apparatus that would scrawl graphics across the screen. At the time that we designed and manufactured this apparatus, computers were much more expensive, much slower, enormous in size when compared with 21st century desktop computers, and had a tiny fraction of the storage capacity of today's mini computers. We had to design and manufacture the compact electronic apparatus needed to drive the dynamic graphics, and to allow the public to press buttons to actively engage in the enrichment activity.

Figure 21-8. Ron Fial calibrates and installs electronics in the visitor response and graphic apparatus.

Figures 21-9a and 21-9b show the final appearance of the device sheltered from the rain in a kiosk overlooking the area where tigers ran to gather meat that was automatically delivered for successful hunts. Graphics scrolled across the screen, thus bringing attention from visitors who had been attracted to the area when they noticed the unusually active tigers. Using electronically controlled dynamic graphics allowed us to easily change the graphics that were seen on the screen as the tigers made progress in learning about their new hunting opportunities. Initially, we programmed the graphic presentations to invite visitors to participate in teaching the tiger about opportunities by pressing buttons to begin specific kinds of hunting opportunities. Later, when the tigers became proficient in pursuit of artificial prey, visitors could assist in randomizing the specific opportunities. During times when no hunting activity was going on, the scrolling graphics told visitors about the natural behaviors of tigers and their conservation status.

Figure 21-9a. Apparatus housed in kiosk.

21-9b. Screen for easy reading of graphics which scrolled to attract the public. Buttons allowed public participation in enrichment activities.

Now that small computers with touch screens have become so reliable, durable, and inexpensive, it would be very cost-effective and practical to use them to allow presentation of dynamic graphics, and allow visitors to participate in enrichment activities by simply touching an icon appearing on the screen.

In addition, there are a number of inexpensive computer software programs for use in easily controlling both audio and visual materials that advance the images and sounds step by step or in a scrolling manner. Almost everyone I know who makes public presentations has one or more of these programs installed in their computer, and regularly uses the software to control the order and timing of audio-visual materials to accompany their talks. An inexpensive computer with one of these software programs can generate wonderful dynamic graphics for educational institutions, such as many modern zoos and museums.

The major defensible reason for continuing to maintain zoos and aquariums in the modern world is conservation education. Finding ways to make graphics really compelling and to continuously evaluate public attention to graphical information should be a principal mission of education departments.

# 22

# CLOSING THOUGHTS

# FROM A STRANGE ANIMAL

Those of you who have no interest in my personal road to a philosophy of enrichment will want to skip this chapter. Should you choose to read it, many of you will find it maudlin in style and/or content. But it is heartfelt, and I wanted to share my feelings with those who are interested as I concluded work on this book. There is only one recipe component in this chapter, but it applies to all enrichment efforts that I personally undertake.

**How People Helped Me Evolve My Thinking and Behavior**

By the time I was twelve I knew that I loved learning about science and philosophy. Before I was twenty I was teaching math in the Air Force, and my thoughts about future goals had begun to fully form. I knew I would want to remain a student for the rest of my life. I hoped it might be possible to have the opportunity to do some useful research, and to share with others any of the knowledge that I acquired. I thought life would be good if I could in any way work toward a career offering me the chance to combine these dreams.

Good luck provided me chances to learn from some brilliant friends and teachers. Alan Watts, who was then headmaster at the Academy of Asian Studies, helped me to recognize clearly that our conception of the "real world" was a product of learned prejudices and our particular sensory abilities. Alan also served as a wonderful teaching model. He always made sure that those he taught came to understand and not just regurgitate what he was trying to convey. As I learned from him about methods of Eastern understanding, it dawned on me that good scientists recognized the always tentative nature of scientific truths and thus had much in common with Zen philosophers. Both emphasized that you could not measure events without the possibility of error and without being influenced by your past history and current surroundings. In later years, I was greatly enriched by occasional opportunities to teach classes in "Eastern Philosophy and Western Scientific Thought" in collaboration with philosophy department professors. I learned much from my colleagues and students in these courses.

Dr. R. W. Porter, who did research at UCLA's Brain Research Institute, arranged a summer position for me as a graduate research anatomist a half century ago. He showed me how to be a good teacher of surgery by setting a great example. His skill in always describing clearly why he was doing each part of a procedure was one I have always tried to emulate. This brilliant man combined a medical career, teaching classes, and conducting research into one package, and still found time to go to baseball games with some of us who were learning from him. My brief experience working under his direction and tutelage reinforced my belief that you could befriend your students without losing their respect. Many of my old students remain very close friends, coauthors, and co-investigators today.

While I was working on my doctorate, wonderful scientists including Nobel laureates were kind enough to allow me to conduct research in their labs and to learn from them. It astonished me that they could make time to learn the details of my research and discuss it with me and with visitors to our lab, including members of congress who influenced the funding of research. Their behavior has doubtless played a role in the fact that I have strived to understand in detail exactly what every person working in my laboratories or field sites is studying, and what they hope to accomplish. This has been an important component in my efforts to foster their success.

Through the years I have been provided grants and contracts to support the work of most of my graduate students as well as myself. Students, often much brighter than me, have brought me joy by choosing to study with me. They have without doubt taught me more than I have taught them as our collaborations progressed. I greatly admire their professional accomplishments and appreciate their continued friendship. Many of the articles, books, and book chapters referenced in this book include these friends as authors or coauthors. It has been a privilege to study along with them, thus helping to fulfill my dream to always remain a student.

I do not know many people who can honestly say that they have been paid to do the work they love best for more than half a century. The gratefulness that I feel for such a rich life continues to motivate me to work. I hope that information, ideas and methods that I have done my best to share in admittedly eccentric and whimsical style will enrich the lives of readers in at least some small way.

For a very long time I have continued to believe that empowering others to have fulfilling lives is the best thing that anyone can do to be a better teacher, parent, or provider of comfort for those in your care. Finding effective means to allow others to gain strength in controlling their own environments, and dealing most effectively with the inevitable contingencies with which life confronts them has greatly enriched my life.

More than two decades of bouts with cancer have taught me some lessons about helping and being helped. I came to realize that I had often failed to include some important considerations in my attempts to enrich the lives of others. Now, I would add emphasis to the fact that one needs to focus on developing the symbiotic relationships required for others to accept your efforts. This is increasingly clear to me in circumstances where I am unable to effectively meet some of life's challenges without assistance. In addition to learning how to relinquish power to others, it is important to provide a means for us animals to accept those things that we cannot independently control. It is important for those whose lives may be improved to learn how to accept help *and* for the caregivers to learn how to make their efforts acceptable for those they wish to help.

I find it fairly easy to accept care from physicians who inform me about my choices as clearly as possible *and* then allow me to make the decisions about what course of action to take in sustaining or ending my life. In contrast, I have great reluctance to simply obey those that adopt a classical parental attitude and assume that I could not possibly know better than they what is good for me. When I am incapacitated and sweet friends and family *offer* assistance rather than insisting that I *must* have assistance, this is much more palatable for me, and I am more inclined to accept their offers. These experiences have led me to increase my efforts to see that when I provide new opportunities for captive animals I make sure that they can choose when and if to use them.

How does all of this relate to the message of this book? It suggests to me that we must focus on accepting guidance from those whose lives we wish to enrich, as well as on finding how to empower them to control more of their own lives. In the case of nonhuman animals this means that despite difficulties imposed by not being able to verbally communicate, we must make every effort to know as much as we can about the individual animals involved. Mastering knowledge about what works in getting others to perform in the manner *you* wish is not the same as providing opportunities for *them* to choose paths that are enriching.

**The "Recipe"**

1. Always begin, whether it is humans or other animals whose lives you hope to enrich, by carefully studying their current behavior and identifying clearly how you can improve their ability to control aspects of their own lives.
2. Be especially careful not to confuse this with a search to find ways to make them behave in the way *you* desire just to bring them into line with your prejudices.
3. Monitor carefully whether the power you have given them to control parts of their environment improves their lives. You may be surprised that their behavior is different from what you anticipated would happen if you relinquished control.

If you work hard at it, empowering others can become contagious. Providing opportunities for animals to become more powerful often has positive consequences for their conspecifics. As individuals hone their skills and become able to live richer lives, others may learn how this came about. We have often seen this in our work. For example, young primates frequently acquired foraging skills by observing their elders feed themselves effectively in enriched captive circumstances. Teaching great apes sign language to help them communicate with us in mutually understandable ways sometimes resulted in *them* teaching those they live with some of these signs and how to effectively use them.

**Effectively Working on Behalf of Animals**

It is inevitable that what I am about to write will be found offensive by some readers. I hasten to assure you that I do not mean to offend or alienate anyone in what follows.

The more we learn about other vertebrates with the ability for cerebral processing, the more long-held beliefs about differences in kind between us and other animals dissolves. Various species have clearly been shown to lie in order to improve their own lot; to use sex to obtain favors from others; to deceive conspecifics in clever ways in order to destroy them; etc. It makes one humble to see how much better other species are at some things than we are.

But there is one area where it remains possible that we have evolved an ability to outdo most other species: extending our love for and work on behalf of others... beyond those efforts that benefit us or those who happen to share our genes.

There are some difficult decisions to be made concerning how to change the behavior of those with whom we seriously disagree. The fact that we are a unique species does not mean that our highly evolved communication and information processing abilities have always served to better life on earth. But I fervently believe that it is immoral to try and change the behavior and beliefs of people by destroying or damaging those with whom we disagree. Physically attacking the families or coworkers of those who treat nonhuman animals in ways that we disapprove will not eradicate their kind of behavior. In fact such acts contradict the purported goals of those who engage in such violent and unkind behavior. One cannot claim to love and be concerned with the wellbeing of *all* animals and at the same time attack other *Homo sapiens*. The only long term road to permanent change beneficial to animals in general is to show by positive example the great benefits that can accrue from living a life in which you strive to improve the lives of all other animals with whom we share this planet.

Humans will doubtless maintain other species in captivity for the foreseeable future, as they have through most of recorded history. It is understandable that some readers of this book may feel that is wrong. There is no question that both human and nonhuman animals are often exploited for selfish interests. While I see merit in actively working to reduce this exploitation and improve the lot of those who are exploited by others, it greatly concerns me that many animal rights activists are paradoxically unkind to human animals. Wouldn't it be more productive and kinder to work to empower captive animals to do some things they enjoy and that benefit their health rather than to terrorize human animals?

I will use a personal example to show that there are difficult decisions to be made in choosing how to better the lives of animals including ourselves and those closest to us. The lives of my daughter and myself have been saved by medical procedures that evolved as a function of research that was done using nonhuman primates to evaluate the effects of these procedures. I am very grateful to those researchers, the animals who served in the research, and medical professionals who have helped to save our lives by using the knowledge developed in the research. Today, I actively work with some of the same kind of biomedical researchers who often join me in efforts to improve the captive lives of the animals used in their current research work.

Many researchers work to reduce the number of animals whose lives are altered in research that benefits humans. Sometimes modern computer models can be employed to evaluate potential outcomes of medical procedures that previously required the use of animals before releasing treatment methods for use on humans. Sometimes artificial devices that process drugs in manners similar to human organs can be employed to minimize the need for testing in nonhuman animals. In my work I encounter a great number of colleagues who unhesitatingly agree that these are salutary advances. Most of those to whom I talk about such things conduct research using animals in their efforts to find ways to cure the diseases of other animals including humans. Their lives are dedicated to finding ways for medical workers and veterinary crews to save the lives of others. And, they respect advances that improve the lives of animals that are used in their efforts.

People employ cute bunnies as companions to please themselves and their children; others use rabbits in medical research, or for testing of some products for safe use by humans. There are bound to be differences between people concerning which, if any, of these practices are necessary and appropriate. But, I do encounter rabbits in all three of the captive circumstances mentioned above. If we care for them as individuals, we should make our best effort to provide them richer captive lives.

Thanks for taking your valuable time to read what I have done my best to share in this book. And thanks to all of you who have been kind enough to empower me by attending my lectures, working with me in the field or lab, sharing your ideas, and reading my sometimes controversial works. You have enriched this old animal's life beyond anything he could have foreseen.

## References


Bayne, K., Mainzer, H., Dexter, S., Campbell, G., Yamada, F., Suomi, S. 1991. The reduction of abnormal behaviors in individually housed rhesus monkeys (Macaca mulatta) with a foraging/grooming board. American Journal of Primatology. 23:23-35.

Bayne, K., Dexter, S., Mainzer, H., McCully, C., Campbell, G., Yamada, F., 1992. The use of artificial turf as a foraging substrate for individually housed rhesus monkeys (Macaca mulatta). Animal Welfare. 1:39-53.

Bennett, E.L., Diamond, M., Krech, D., Rosenzweig, M.R. 1964. Chemical and Anatomical Plasticity of Brain: Changes in brain through experience, demanded by learning theories, are found in experiments with rats. Science. 146:610-619.

Bruce, H.M. 1963. Olfactory block to pregnancy among grouped mice. J. Reprod. Fertil. 6:451.

Burr, L. 1997. Reptile Enrichment: Scenting for Response. Shape of Enrichment. 6.

Cheney, C.D. 1978. Predator-Prey Interactions. In Behavior of Captive Wild Animals. H. Markowitz, Stevens, V., (eds). Nelson-Hall, Chicago. 1-19.

Chivers, D. 1972. The siamang and gibbon in the Malay Peninsula. In Gibbon and Siamang. Vol. 1. D. Rumbaugh, (ed). S. Karger, Basal. 103-135.

Coe, J.C. 1989. Naturalizing habitats for captive primates. Zoo Biology Supplement:117-126.

Cornick, L.A., Markowitz, H. 2002. Diurnal Vocal Patterns of the Black Howler Monkey (Alouatta pigra) at Lamanai, Belize. Journal of Mammalogy. 83:159-166.



Cornick, L.A., Markowitz, H. 2002. Diurnal Vocal Patterns of the Black Howler Monkey (Alouatta pigra) at Lamanai, Belize. Journal of Mammalogy. 83:159-166.



Erwin, J., Deni, R. 1979. Strangers in a strange land: Abnormal behaviors or abnormal environments? In Captivity and Behavior. Primates in Breeding Colonies. Laboratories and Zoos. J. Erwin, Maple, T.L., Mitchell, G., (eds). van Nostrand Reinhold, New York. 1-28.

Forthman-Quick, D.L. 1984. An integrative approach to environmental engineering in zoos. Zoo Biology:29-43.

Foster-Turley, P., Markowitz, H. 1982. A captive behavioral enrichment study with Asian small-clawed river otters (Aonyx cinerea). Zoo Biology. 1:29-43.

Fradrich, H., Lang, E. M. 1972. Hippopotamuses. In Animal Life Encyclopedia. Vol. 13. B. Grizmek, (ed). Van Nostrand Reinhold, Zurich. 109-129.

Gavazzi, A.J., Cornick, L.A., Markowitz, T.M., Green, D., Markowitz, H. 2008. Density, Distribution, and Home Range of the Black Howler Monkey (Alouatta pigra) at Lamanai, Belize. Journal of Mammalogy. 89:1105-1112.

Gibson, K. 1999. Personal Communication.

Grams, K., Ziegler, G. 1995. Animal Keeper's Forum.

Grigg, E. and Markowitz H. 1997. Habitat use by bottlenose dolphins (Tursiops truncatus) at Turneffe Atoll, Belize. Aquatic Mammals. 23:163-170.

Grzimek, B. 1972. Camels and Llamas. In Grzimek's Animal Life Encyclopedia. Vol. 13. B. Grzimek, (ed). Van Nostrand Reinhold, New York. 130-148.

Gustavson, C.R., Brett, L.P., Garcia, J., Kelly, D.J. 1978. A Working Model and Experimantal Solutions to the Control of Predatory Behavior. In Behavior of Captive Wild Animals. H. Markowitz, Stevens, V., (eds). Nelson-Hall, Chicago. 21-66.

Hediger, H. 1955. Studies of the Psychology and Behaviour of Captive Animals in Zoos and Circuses. Butterworth's Scientific Publications, London.

Khan, C.B., Markowitz, H., McCowan, B. 2006. Vocal development in captive harbor seal pups, Phoca vitulina richardii: Age, sex, and individual differences). Acoustical Society of America. 120:1684-1694.

Langbauer, W.J., Payne, K., Charif, R. and Thomas, E. 1989. Responses of captive elephants to playback of low-frequency calls. Canadian Journal of Zoology. 67:2604-2607.

Lehner, P.N. 1979. Handbook of Ethological Methods, First edition. Garland STPM Press, New York. 403 pp.

Lehner, P.N. 1996. Handbook of Ethological Methods, Second edition. Cambridge University Press, Cambridge. 672 pp.

Leveille, J., Wright, M., Edwards, S. A. 1993. A sandbox and Bungee Boots for Hosenose. *The Shape of Enrichment* 2(3):1-2.

Line, S.W., Morgan, K., Markowitz, H., Strong, S. 1989. Evaluation of attempts to enrich the environment of singly-caged non-human primates. In Animal Care and Use in Behavioral Research: Regulations, Issues, and Applications. J.W. Driscoll, (ed). Animal Welfare Information Center, National Agricultural Library, Beltsville. 103-117.

Line, S.W.,Morgan, K., Markowitz, H., Strong, S. 1989. Influence of cage size on heart rate and behavior in rhesus monkeys. American Journal of Veterinary Research. 50:1523-1526.

Line, S.W., Clark, A.S., Markowitz, H. 1989. Responses of adult female rhesus macaques to nylaballs and manipulable novel objects. Lab Animal. 18:33-40.

Line, S.W.,Clark A.S., Markowitz, H., Ellman, G. 1990. Responses of female rhesus macaques to an environmental enrichment apparatus. Laboratory Animals. 24.

Line, S.W.,Morgan, K.N., Roberts, J.A., Markowitz, H. 1990. Resocialization on health and behavior of aged rhesus macaques (Macaca mulatta). American Journal of Primatology. 29:8-12.

Line, S.W., Markowitz, H., Morgan, K., Strong, S. 1991. Cage size and environmental enrichment: effects upon behavioral and physiological responses to the stress of daily events. In Through the Looking Glass: Issues of Psychological Well-being in Captive Non-Human Primates. M.A. Novak, Petto, A., (ed). American Psychological Association, Washington. 160-180.

Line, S.W., Morgan, K.N., Markowitz, H. 1991. Simple toys do not alter the behavior of aged rhesus monkeys. Zoo Biology. 10:473-484.

Maple, T.L. 1980. Orangutan Behavior. Van Nostrand Reinhold, New York.

Maple, T.L., Hoff, M.P. 1982. Gorilla Behavior. Van Nostrand Reinhold, New York.

Markowitz, H., Sorrells, J. 1969. Performance of "maze-bright" and "maze-dull" rats on an automated visual discrimination task. Psychonomic Science. 15:357-358.

Markowitz, H., Becker, C.J. 1969. Superiority of "maze-dull" animals on visual tasks in an automated maze. Psychonomic Science. 17:171-172.

Markowitz, H. 1972. A Report. Oregon Zoological Research Center. 11-15.

Markowitz, H., Schmidt, M., Nadal L., Squier, L. 1975. Do elephants ever forget? Journal of Applied Behavior Analysis. 8:333-335.

Markowitz, H., Stevens, V. 1978 (eds). The Behavior of Captive Wild Animals. Nelson Hall, Chicago.

Markowitz, H. 1982. Behavioral Enrichment in the Zoo. Van Nostrand Reinhold, New York.

Markowitz, H., Spinelli, J. S. 1986. Environmental Engineering for Primates. In Primates: The Road to Self-Sustaining Populations. K. Benirschke, editor. Springer-Verlag, New York. 489-498.

Markowitz, H., LaForse, S. 1987. Artificial prey as a behavioral enrichment for felines. Applied Animal Behavior Science. 18:31-43.

Markowitz, H., Line, S.W. 1989. Primate research model and environmental enrichment. In Housing, Care and Psychological Well-Being for Laboratory Primates. E. Segal, (ed). Noyes, Park Ridge. 203-212.

Markowitz, H. 1990. Environmental opportunities and health care. In Handbook of Marine Mammal Medicine: Health Disease and Rehabilitation. L. Dierauf, (ed). CRC Press, Boca Raton. 483-488.

Markowitz, H., Aday, C., Gavazzi, A. 1995. Effectiveness of acoustic "prey": Environmental enrichment for a captive African leopard (Panthera pardus). Zoo Biology. 14:371-379.

Markowitz, H., Gavazzi, A.J. 1996. Definitions and goals of enrichment. In The Well-Being of Animals in Zoo and Aquarium Sponsored Research. G. Burghardt, Belitzki J., Boyce D, et. al, (eds). Scientists Center for Animal Welfare, Greenbelt, MD. 85-90

Markowitz, H. 1998. The conservation of species-typical behaviors. Zoo Biology. 16:1-2.

Markowitz, H., Eckert, K. 2005. Giving Power to Animals. In Mental Health and Well-Being in Animals. F.D. McMillan, (ed). Blackwell Publishing, Ames. 201-209.

Markowitz, H., Timmel, G.B. 2005. Animal Well-Being and Research Outcomes. In Mental Health and Well-Being in Animals. F.D. McMillan, (ed). Blackwell Publishing, Ames. 277-283.

McPhee, M. 2002. Intact carcasses as enrichment for large felids: Effects on on- and off-exhibit behaviors. Zoo Biology. 21:37-47.

Mellen, J.D., Stevens, V.J., Markowitz, H. 1981. Environmental enrichment for servals, Indian elephants and Canadian otters. In International Zoo Yearbook. Vol. 21. Zoological Society of London, London. 196-201.

Melo, L. 1999. Auditory Enrichment for Asian Elephants. The Shape of Enrichment. 8:1-4.

Neuringer, A. 1969. Animals respond for food in the presence of free food. Science. 166:399-401.

Redford, K.H. 1985. Feeding and food preferences in captive and wild giant anteaters *(Myrmecophaga tridactyla).* Journal of Zoology Series A 205:559-572.

Saskia, J., Schmid, H. 2002. Effect of feeding boxes on the behavior of stereotyping amur tigers (Panthera tigris altaica) in the Zurich Zoo, Zurich, Switzerland. Zoo Biology. 21:573-584.

Schanberger, A. 1991. Enrichment techniques for elephants at the Phoenix zoo. In American Association of Zoological Parks and Aquariums, 1991 Regional Conference Proceedings.

Schmidt, M., Markowitz, H. 1977. Behavioral engineering as an aid in the maintenance of healthy zoo animals. Journal of the American Veterinary Medical Association. 171:966-969.

Schuett, E.B., Frase, B. A. 2001. Making scents: using the olfactory senses for lion enrichment. The Shape of Enrichment. 10.

Seyfarth, R., Cheney, D. 1980. Vocal recognition in free-ranging vervet monkeys. Animal Behavior. 28:362-367.

Shaw, J.H., Mchad-Neto, J., Carter, T.S. 1987. Behavior of Free-living Giant Anteaters *(Myrmecophaga tridactyla). Biotropica.*19(3:):255-259.

Stevens, V.J. 1978. Basic Operant Research in the Zoo. In Behavior of Captive Wild Animals. H. Markowitz, Stevens, V.,( eds). Nelson-Hall Inc., Chicago.

Tapp, J.T., Markowitz, H. 1963. Infant Handling: Effects on avoidance learning, brain weight and cholinesterase activity. Science. 140:486-487.

Tolman, E.C. 1924. The inheritance of maze-learning ability in rats. Journal of Comparative Psychology.

Tryon, R.C. 1940. Genetic differences in maze-learning ability in rats. Yearbook of the National Society for the Study of Education. 39:111-119.

Yanofsky, R., Markowitz, H. 1978. Changes in general behavior of two mandrills (*Papio sphinx*) concomitant with behavioral testing in the zoo. Psychological Record. 28:369-373.

Ziegler, G., Roletto, J. 2000. Enrichment options: the power of scent. Animal Keeper's Forum. 27:355-357.

# Index

## D



## E



## F



## G



## H



## I

## J



## K



## L



## M



## N



## O



## P

## R



## S



## T



## U



## W



## X



## Y



## Z

CPSIA information can be obtained at www.ICGtesting.com
Printed in the USA
LVOW031529290112

265071LV00001BA/2/P

9 780983 357919